Slaves of Mercury

by

Nathan Schachner

Double9 BOOKS

Slaves of Mercury

by Nathan Schachner

ISBN: 978-93-59328-65-2

Published by

DOUBLE 9 BOOKS
2/13-B, Ansari Road
Daryaganj, New Delhi – 110002
info@double9books.com
www.double9books.com
Tel. 011-40042856

ABOUT THE AUTHOR

Nathaniel Schachner (January 16, 1895 – October 2, 1955) was an American writer, historian, and attorney who advocated for the creation of rockets for space travel. He authored under the pen names Nat Schachner and Nathan Schachner. Schachner, a well-known author of historical books on people from America's Revolutionary War, was also a regular contributor to the genre in the years leading up to and during what became known as the Golden Age of Science Fiction (c. 1938-1946). Schachner is best known for biographies of American historical leaders such as Thomas Jefferson and Alexander Hamilton, although he began his literary career by contributing short tales to renowned "pulp magazines" specializing in science fiction, horror, mystery, and adventure. He produced almost fifty pieces to periodicals such as Astounding pieces, Terror Tales, Horror Stories, Dime Mystery Magazine, and Fantastic Adventures during the Great Depression. He then began writing historical nonfiction and fiction, receiving acclaim for his exhaustive research.

Schachner, a lawyer, was a founding member and officer of the American Interplanetary Society, which pioneered liquid fuel rocketry in the United States in the early 1930s.

CONTENTS

CHAPTER I
THE SPACE WANDERER RETURNS

Hilary Grendon piloted his battered, time-worn space flier, the *Vagabond*, to the smiling Earth that rose rapidly to greet it. Only the instinctive ease of long practise prevented a smash-up, his hands trembled so at the controls.

Hilary returns to find alien diskoids in Earth's stratosphere, and out-world lords patrolling her cities.

Home again—the old familiar Earth! He could scarcely believe it! Perhaps it was only a dream, and he'd wake up among the unhuman glittering cylinders of Saturn, shuddering and crawling with the iciness of their fixed regard.

Hilary's eyes blurred with unaccustomed mistiness as he drank in the warm sunlight, the soft green of the grass and the gracious lines of the slender birches as they fluttered their leaves daintily in the unhurrying breeze. How different it all was from the harsh red angularities of Mars!

He was outside, breathing deeply, inhaling the perfumed air with delight. This was the only heaven; beyond—that far-flung immensity of planetary orbs—was hell! He, Hilary Grendon, the carefree, smiling skeptic of old, was a Fundamentalist now.

How long was it since they had started out on the first flight that man had taken into outer space—he and those stanch comrades? Five years? God! Had it been so long? Yet here he was, back on Earth again, the kindly, blessed Earth their eyes had clung to when they were fighting desperately for their lives against the protoplasmic things that inhabited Ganymede.

Hilary brushed a tear away as he thought of those brave, loyal friends. Dick lay as he fell on Saturn, transfixed by an icicle dart; Martin had been engulfed in an unholy maw on Ganymede; Dorn was a frozen idol to the spiral beings of Pluto; and poor Hurley, his fate was the worst of all: his hideously bloated body was swinging in an orbit around Mars, a satellite through all eternity.

He, Hilary Grendon, was the sole survivor of that tremendous Odyssey!

Hilary shook his head vigorously to clear away the flood of recollections. Enough that he had returned. Then a sudden eagerness surged through him, a joyous intensity of emotion. What a story he had to relate—how the Earth people would hang with bated breath upon his adventurings! And Joan—his heart gave a queer leap at the thought of that slender ardent wisp of a girl with her shining head and steady gray eyes. She had promised to wait for him, forever, if need be. She had said it simply, without heroics; yet Hilary knew then that she would keep her promise.

A rush of impatience succeeded the inaction of his memories. He must get to New York at once. He could not wait any longer. Joan first—then Amos Peabody, the venerable President of the United States—to report his return. He smiled at the stupefaction that would greet him. No doubt he had long been given up for dead. The world had been skeptical of the space ship he had invented; had, except for a faithful few, mocked at his plans. Indignantly he had taken his calculations, his blue prints of the spheroid, along with him. If the flight was a success, well and good; if not, they would not be worth much anyway.

In spite of his fever to be off, he carefully locked the controls, sealed the outer air-lock. Hilary Grendon was a methodical man: that was the reason he had survived.

Then he struck across country, walking very fast. He knew where he was: in the wilderness of the Ramapos, some forty miles from New York. Sooner or later, he reasoned, he would strike one of the radiating conveyors that led into the metropolis, or a human being that would set him on the right track.

A half hour's sturdy tramping brought him out of the tangled hills into civilization. There was a glitter of metal and vita-crystal dwellings that stood four-square to the sun and the winds. A broad ribbon-conveyor hurled its shining length in ceaseless rush down the narrow valley. Human beings—normal homely Earth men with the ordinary number of legs and arms, with honest-to-God faces and warm living flesh, were seated on the conveyor-benches as they flashed by. Hilary could have wept with delight. It was two years since he had seen his own kind; two years since Hurley's tragic misstep through the breach in the air-lock made by a meteor as they were nearing Mars.

Hilary leaped on the slow-moving ramp, skilfully worked his way across the graded speed belts until he was on the express conveyor that led straight on to New York.

He sank into a cushioned seat next to an oldish man with iron-gray hair through which the speed of their flight whipped and pulled. Hilary was bursting for real human conversation again; he grinned to himself at the excited astonishment of this impassive stranger if he should announce himself. How should he do it? Should he remark casually without any preamble: "Pardon me for addressing you, sir, but I'm Hilary Grendon, you know." Just like that, and lean back for the inevitable gasp: "What, not *the* Hilary Grendon!" And he would nod offhandedly as though he had just taken a little trip to Frisco and back.

He stole a sidelong glance at the sternly-etched profile. The man was staring straight in front of him, looking neither to the left nor to the right. It did not seem as if he were aware of Hilary's existence. So with a sigh Hilary decided against that method of approach as a trifle too abrupt.

"Nice day to-day, isn't it?" The sound of his own voice startled him. English was an alien language to his unaccustomed tongue after the hissing syllables of the Martians.

With pathetic eagerness he awaited the inevitable answer to this commonplace introduction; that he might once more hear normal Earth tones in friendly converse, see the smile of greeting on a real Earth face.

But there came no answer. The man continued staring straight ahead, immobile, fixed. There was no slightest turn to the etched profile. It was as if he had not heard.

Hilary felt a sudden surge of anger. Damn discourteous, this first Earthman he had met. What had happened to the old hospitality? Had it passed out while he was roaming the spaces? He leaned over, harsh words tumbling for exit, when suddenly he checked himself. There was something strange about that fierce blank stare. The man's face, too, he saw now, was lined and worn; suffering had left its multitudinous imprint upon an ordinarily rotund countenance.

Hilary shouted suddenly: "Good morning." The man did not answer, nor did he stir from his unvarying pose. Deaf! The returned Earthman suffered swift pity. With gentle forefinger he prodded the man.

The reaction was astounding. The man cowered like a pricked balloon; little strangling moans forced themselves out of clenched teeth. Dumb, too! His face jerked around to the direction of Hilary's gentle prodding. Merciful heavens, the man was blind also! Two vacant red-rimmed sockets stared pitifully at him—the eyeballs were gone, ripped out.

But what struck Hilary particularly was the mortal terror that was depicted on the blind man's face. It was as though he expected some cruel, crippling blow to follow; as though it were the last straw on the back of unmentionable former agonies. Hilary shuddered. It was not good to witness such animal fear. A dark shadow blotted out the brightness of the Earth-day for him. There was something wrong here, something that required a good deal of explanation.

His probing eyes went thoughtfully over the poor cowering wretch. Those careworn features were familiar, somehow. Where had he seen the man before? Suddenly he stiffened, choking an exclamation. The man was bound immovably to his seat. Thin metal links, almost invisible, encircled his feet; held the elbows taut against the fluted columns of the seat-back.

Hilary's space-tanned features hardened; the light gray of his eyes darkened. All the pleasure of his homecoming vanished. The kindly Earth seemed suddenly grown inimical. What had happened in the five long years of his absence? This would have been impossible on the Earth he had known; a man, manifestly the victim of hideous tortures, bound like a wild animal to the seat of a public conveyor.

He went swiftly into action. From the depths of a capacious pocket he fished a sheathed blade of stellite, triply keen; its razor-sharp edge sawed smoothly at the bonds.

In his mounting anger Hilary had paid no attention to the scattering of people occupying the cushioned chairs of the speeding conveyor. But a smothered nearby gasp caused his head to jerk up. He met the incredulous stare of a paunchy, heavy-jowled man seated some chairs away. There was more than incredulity, there was furtive fear in the small beady eyes sunken in folds of fat.

Hilary gave way to unreasoning anger.

"Stop looking like a stuck pig," he called sharply. "Give me a hand with this poor fellow."

The response was surprising. The man got up from his chair precipitately, stark panic written all over him. The sweat oozed from his shiny forehead as he backed cautiously away. He tripped over the edge of the seat behind, and fell. Once more he scrambled to his feet, and as if the fall had released his trembling muscles, he turned and ran, stumbling and dodging across the local conveyors, never once looking back.

Hilary watched his mad flight wonderingly. "Good Lord," he thought, "does my face frighten people so? Maybe I've turned into a Martian."

He turned to appeal to the others on the conveyor, and received another shock. The few men within earshot were already on their feet and moving away from there with unostentatious celerity. Hilary surveyed their receding backs thoughtfully. What was there about himself to frighten grown men out of their wits? Or was it the poor tortured wretch he was trying to release who was responsible for the exodus?

Already the express was almost clear. He saw the deserters throwing themselves guiltily into seats on the local belts, and then he was carried swiftly past. Only one man remained stubbornly in his seat, some fifteen rows back. He was a huge mountain of a man, a giant upon Earth, and there was a strangeness in his wide stare.

Hilary frowned, then shook his head, and dropped down to his task again. The blind man moaned and jerked as he felt the bite of stellite upon his fetters. Hilary made soothing sounds, forgetful that he could not hear, and worked steadily. There was a little clinking noise and the links that bound the arms fell apart. He attacked the leg shackles next.

There was a tap on Hilary's shoulder, light, electric, yet strangely heavy in its implications. Hilary turned his head sharply, saw the landscape blotted out by a huge overshadowing bulk. Five years in a hostile universe had made him cautious. He pivoted on his heels and rose in a single flowing motion, stellite blade ready for instant action.

CHAPTER II
THE STRANGE GUARD

There confronted him the hugest figure of a man he had ever seen. Hilary was not lacking in inches himself—he was well over six feet; but the giant staring quizzically down at him was nearer seven, with shoulders to match. The features of his face were gargantuan in their ruggedness, yet singularly open, while a pair of mild blue eyes, childlike in expression, looked in perpetual wonder out upon the world.

In spite of his annoyance, Hilary instinctively liked the giant.

"What do you want?" he inquired gruffly.

The Colossus surveyed him with his child's eyes.

"Man, you are crazy." He spoke in a deep bass rumble, without emotion or inflection. He was simply stating a fact.

A surge of annoyance swept over the returned wanderer from the far spaces. This was the last straw.

"I may be," he admitted coldly, "but I like my particular form of craziness."

"You know the penalty of course for what you are doing?" the big man inquired unemotionally.

Hilary swore deeply. "Damn the penalties, whatever you mean by that. Here's a man who has been tortured unmercifully—chained like a dog. I intend to free him."

The mild blue eyes contained the hint of a gleam.

"But you know the penalties," he repeated. His murmur sounded like the rumble of a distant earthquake.

Hilary straightened sharply, poked his finger at the midriff of the giant.

"I don't know what you are talking about," he stabbed. "What is the meaning of all this? Who is this unfortunate, and why did everyone disappear as though I had the plague when I sat next to him?"

A look of bewilderment swept over the massive face, bewilderment tinged with a dawning suspicion of the questioner's sanity.

"You mean to say you don't know?" The tone held incredulity.

"I've just told you so," Hilary pointed out. He felt a growing unease.

The giant eyed him closely. "Man, where on earth have you been these last three years?"

Hilary grinned. "I haven't."

"You haven't?" echoed the other. Suspicion hardened the childlike eyes into cold flame. The man was dangerous when aroused. He thrust his jaw down at Hilary. "If you are jesting with me...." He left the sentence unfinished, but the clenching of a huge fist left no doubt as to his intention.

"I am not jesting," Hilary assured him grimly. "I have been away from the Earth for five years. I've just returned."

The great hand clenched tighter. "Now I know you are crazy, or—Who are you?" he ended abruptly.

"Hilary Grendon."

"Hilary Grendon—Hilary Grendon," rumbled the other in manifest perplexity. It was evident the name meant nothing to him.

This then was the homecoming he had dreamed of in the unfathomable reaches of space. Hilary thought bitterly. Five short years and he was already forgotten. Then the irony of it struck him, and he laughed aloud.

"Yes," he said. "Five years ago I led the Grendon Expedition to explore interplanetary space in the space-ship I had invented. I've come back—alone."

It was amazing to watch long-overlaid memories struggling up through the subconscious. At last the giant spoke.

"Oh, yes," he said meditatively, "I seem to remember something about it." He surveyed Hilary with a new interest. "So you were one of those chaps, eh?"

The explorer admitted it, humbly. Of such are the uses of fame.

"Well, now," said the giant, "that might explain it. Though it sure beats all." And he shook his head as though he still did not understand.

"Who is that man?" Hilary stabbed a forefinger at the blind man, who sat immobile as before, his worn etched face ever to the front. "It's monstrous. Amos Peabody shall hear of it."

The Colossus looked at him mildly.

"That," he said, "*is* Amos Peabody!"

Silence lay like a live thing between them. Hilary whirled in a kaleidoscope of emotion. Was this wasted, tortured being the portly, dignified President of the United States who had bade him Godspeed at the start of his tremendous journey five years before? His pitying eyes searched the lineaments of the poor wretch. There was no doubt of it now; it *was* Amos Peabody.

Hilary gripped his informant's arm. His voice was deadly calm. "I want the truth about this, and I want it fast."

"The truth," echoed the big man with strange laughter; "now that is something—"

His eyes widened over Hilary's shoulder. With a swiftness remarkable in one of his bulk he shook off Hilary's restraining grip, caught him by the shoulder and thrust him, all in one motion, into a chair several removed from Peabody. In a trice his huge bulk was safely ensconced in the adjoining one.

Hilary's hand went to the butt of the automatic within his blouse. The giant saw the movement. He leaned forward.

"Don't make a move," he warned, "the guard is coming."

"What guard?"

"You'll see fast enough. Appear unconcerned if you value your life. Don't look back."

Hilary complied. His face became an expressionless mask as he lounged in his chair, but his thoughts seethed and boiled. What terrible mystery had enveloped the Earth during his absence? Why was Amos Peabody tortured and made into a public mockery?

There was a slight whirring noise behind him. Heedful of his companion's admonition he relaxed in apparent unconcern, but his hand stole once more to the fold in his blouse. His long fingers rested caressingly upon the butt of his automatic. There were still three good Earth bullets in the chamber.

The whirring ceased. There was a slight jar as of something landing on the speeding conveyor. Yet Hilary did not look back, though his grip tightened. A heavy body stumbled toward them, cursing in strange phrases. It passed from behind, came to a halt before the giant. Hilary shot a sharp glance upward from under veiled lids. An exclamation sprang full-throated to his lips, died unheard under a tremendous effort of his will.

Before them stood a being—it could not be called a man. He was no denizen of the Earth, that was evident, yet Hilary had visited all the planets outward from our own without encountering such a monster.

He hulked before them like a behemoth, even dwarfing Hilary's companion with his enormous stature; but it was noticeable that he supported his weight ill, as if Earth's gravitation was too strong for him. Manlike he was in every essential, but the skin of his face was a pasty dull gray, and ridged and furrowed with warty excrescences. Two enormous pink eyes, unlidded, but capable of being sheathed with a filmy membrane, stared down at them with manifest suspicion. A gray, three-fingered hand held an angled tube significantly. A lens gleamed transparent in the sunlight from the open end.

Hilary did not move under the stare, nor did his companion. The mild blue eyes were childlike as ever. The guard's gaze shifted from them to the trembling figure of Amos Peabody. He bent over him, thrust at him with ungentle hand. The automatic under Hilary's fingers crept farther out from the blouse, but a warning gesture from his companion stopped him.

The guard amused himself with shaking the blind man; then he bent suddenly. He had seen the broken links. With ominous deliberation he turned his vast weight upon them. His baleful pink eyes fastened upon Hilary's companion.

"You!" he growled throatily, "what do you know about this?" He spoke in English, but it was obviously not his native tongue.

Mildly innocent was the giant's face.

"I know nothing, Magnificent," he said humbly. "I am on my way to Great New York on my own insignificant affairs, and I bother my head with nothing else."

"The bonds of this dog, Peabody, have been severed," the guard insisted, "and recently, too. Speak up, Earthman, or—you know the penalty."

"I know the penalty," he answered respectfully, "but I have been seated here only five minutes, and I know nothing of this Peabody."

The guard fingered his tube.

"Let me see your tag," he said suddenly.

The other opened his blouse obediently and exposed a thin copper disk suspended on his chest. The guard tugged at it brutally to bring it within range of his vision. The pull jerked the giant's head forward, and the thin metal strand cut cruelly into the back of his neck. Hilary saw a flush of red sweep like a wave up to his forehead, and the mild blue eyes turned hard like glinting blue pebbles. But not a word escaped his lips.

"Grim Morgan," the guard read, "A46823 Great New York. Pah, what barbarous names you Earthmen have." He shoved the giant back heavily into his seat, and turned his baleful glare upon Hilary.

"You, what do you know about this?"

Grim Morgan interposed hastily. "Nothing, Magnificent. He came on the express conveyor after I did."

The guard's free hand went back. Very deliberately he struck him across the face with three ridged fingers. An angry welt raised.

"That will teach you to keep your mouth shut when not spoken to."

The big man's eyes were mild, but his hands tensed as though they were curled around a throat. He said nothing.

The guard turned to Hilary again. "Answer me," he barked.

"My friend told the truth," Grendon said simply.

"Your tag?"

"I have none."

Suspicion flared openly in the pink eyes.

"Where is it?"

"I never had one."

"Ah!" There was a world of meaning to the exhalation. "You know of course that every Earthman must be registered. The penalty for non-obedience is—death."

The angled tube came up with the swiftness of light. Grim Morgan cried out sharply, lunged out of his seat. Hilary tore at his gun, knowing sickeningly that the draw would be slower than the action of the strange weapon in the guard's hand.

There was a sneer on the monster as he pressed something on the tube. Hilary's automatic was only half out of his blouse. Grim's lunge would never reach in time. He was too far away.

CHAPTER III
THE DEATH OF AMOS PEABODY

Just how any inkling of what was happening penetrated the pain-swept consciousness of the blind and deaf President could never be determined. Possibly a thin repercussion of Grim's cry, possibly an intuition that comes to sense-bereft men. But he had jerked spasmodically erect. There was a sharp tinkling as the weakened leg links broke. He threw himself in a queer, awkward movement forward, directly in the path of the tubed weapon. A blinding beam flashed out of the orifice, sheared through Peabody's middle as though he were cut cleanly in half with a gigantic knife. He toppled in two sections to the floor of the conveyor—released from all humiliation, all suffering.

At the same time two other things happened. Grim Morgan hit the guard like a crashing thunderbolt and Hilary's gun barked once. The monster tottered under the impact. A puzzled expression flitted over his pinkish eyes, a filmy sheath spread over them like a veil, and he fell heavily, a neat bullet hole square between his eyes.

Hilary shoved the gun back in his blouse, and stared alternately at the huddled form of the grotesque being and all that remained of Amos Peabody. The old President had saved his life at the cost of his own. Instinctively his hand went up in formal salute to the gallant old man.

Grim Morgan shook him by the shoulder.

"Man," he said quietly, "we have killed a Mercutian guard. Within the hour we shall be dead men too."

Hilary looked up at him sharply.

"A Mercutian," he echoed. "You mean—"

"That for three years now the Earth has been a conquered province of these devils from Mercury," Grim interposed swiftly. "We have committed the unforgivable offense and must pay for it."

Hilary glanced swiftly around. The express conveyor was clear of passengers for over a hundred yards each way. All the people within range had cleared off when Hilary had attempted to release Peabody. The small figure of a man got up from his chair beyond the charmed circle, and was threading his way forward. The local conveyors seemed to be moving backward at graded speeds. Beyond was the open country, gradually thickening into scattered rows of crystal buildings. They were in the suburbs of Great New York. Within ten minutes the conveyor terminal would be reached.

Hilary's eyes flicked speculatively to the tiny cigar-shaped boat in which the dead guard had flown down to them. Its smooth gray-gleaming surface was devoid of wings or other lifting devices. Only a fan-shaped fin projected from the stern like the tail of a fish. The cockpit, if such it could be called, was tiny, just ample enough to accommodate the Mercutian's girth. The sunlight dazzled back from a bewildering jumble of tiny lenses inset in the instrument board. Arranged along the hull, on either side, were larger disks of the same quartz-like material.

"Let's get away in the flier," he said.

"Can't," Grim said. "Those lenses you see on the instrument board are the controls. No one knows how to operate them except the Mercutians. Our people managed to capture a few, but couldn't do a thing with them."

Hilary stared at the motionless flier with interest. "What are those round glass disks stretched along the hull in a double row?" he asked. "They look like burning glasses."

"That's just what they are," said Grim sadly. "The top row are sun-lenses, that throw a terrible ray for a distance of two to three hundred feet. Melts everything in its path—men trees, rocks even. You saw one in action in the sun-tube with which poor old Peabody was cut in half. The lower row of lenses on the flier are search beams."

"Search beams?" Hilary echoed inquiringly.

"Yes. They act like X-rays, more powerful though, and with the further property of rendering everything they touch transparently crystal for depths of ten to fifteen feet. Lead is the only element they can not penetrate. Another secret our scientists can not fathom, so they talk learnedly about the stream of rays polarizing the structure of matter along a uniaxis."

"Can't those lenses be duplicated, and turned as weapons against the Mercutians?"

"No. They are made of a peculiar vitreous material native to Mercury."

"And no one has found out the principle on which they work?"

"Well, there have been theories. We haven't many scientists left, you know. But the most popular one is that these lenses have the power of concentrating the rays of the sun to an almost infinite degree, and then spreading them out again, each individual beam with the concentrated energy of the whole. Some new way of rearranging quanta of energy."

"Hmm!" Hilary's brow was wrinkled. For a long moment he stared and thought.

At last he snapped back to their present situation: the dead guard at their feet, the dismembered body of Amos Peabody, the cowed groups of Earthmen on the speeding conveyors, keeping respectful distances.

"We'd better start moving if we want to get away," he said.

"It's no use." Grim spread his hands resignedly. "We'll have to take our medicine."

Hilary flared angrily. "You're talking nonsense. What's to prevent us from hopping to another platform? There is no other Mercutian in sight."

"No, but there were plenty of Earthmen who saw us."

"They won't tell."

"Oh, won't they?" Grim shook his head quietly. "You don't realize what has happened. Their spirit has been crushed until they are actually slavish in soul as well as in body. They fought bravely enough on the first invasion. Even after the conquest there were plenty of men looking for an opportunity to fight them again. Amos Peabody headed the revolt. It was smothered in blood, so effectually that only slaves are left. Peabody was left as a horrible warning. He was sent from city to city to be exhibited to the populace, unattended on the way, so confident were the Mercutians of the terror they had inspired."

"So you think those Earthmen who saw us will report to their masters," Hilary said slowly.

Grim nodded.

"I know it—they'll expect to curry favor in return."

Hilary felt a web of circumstance tighten around him. His jaw tautened. Thank the Lord he had been away—on his own. He had not the soul of a slave—yet.

"Won't you fight for your life?" he asked the big man curiously.

A spark lit in the mild blue eyes, died down.

"Yes if there were a chance," he said dully. "But there is none. The whole Earth is honeycombed with their guards. They have fliers, sun weapons, invisible search beams. We'd never elude them."

Hilary snorted impatiently. "We have good Earth brains, haven't we? I've traveled all the outer planets and never met any intelligence equal to that of a man, and I won't admit for a moment that the Mercutians are any exceptions."

A man stepped casually onto the express, took one startled look at the dead guard, at them, and fled precipitately back.

Another one to spread the alarm," Morgan said grimly. "There'll be a dozen guards dropping down on us in the next five minutes."

"Let's get going then." Hilary was pulling the big man along by main force when he heard a movement in back of them. He stopped, whirled, automatic thrusting its blue nose forward.

The little man who had gotten up before on the express was pushing rapidly toward them.

"Stop." Hilary's voice was harsh with command.

But the little man did not heed. He literally stumbled in his haste, crying: "You've killed a Mercutian."

"What of it, my bantam?" Hilary inquired softly, the muzzle of his gun boring into a lean flat stomach. The little man was actually pressing against the automatic in his excitement.

"What of it?" he shrilled excitedly. "God, all this time I've been waiting to find someone with guts enough to smash one of them. Sir, I'm proud to shake your hand."

He reached over the wicked-looking muzzle, gripped Hilary's fist, still tight on the gun butt, and pumped vigorously. He dropped the hand, swerved on Grim.

"And you too, sir." His little fingers were engulfed in a mighty paw. "I saw it all, I tell you," he babbled. "We've got them on the run. We'll sweep the filthy devils clean off the Earth. We'll annihilate them."

"Whoa there, my little gamecock." Morgan grinned down at the excited little man. "One Mercutian doesn't make a Roman holiday. They're plenty more where he came from. You'd better clear out before they come, or you'll be included in the party."

The little fellow—he was not much more than five feet no inches tall—drew himself up to his full height. "What," he ejaculated, "me desert my friends? Wat Tyler's never had that said of him yet. We stick together, to hell and back again."

Hilary grinned as he slipped the weapon back into his blouse. He was beginning to like this little firebrand. In truth, Grim had rather fairly described him as a gamecock. His stature, the bristly red hair that flamed above a freckled face, the lightest of blue eyes that snapped with excitement, the peculiar strut of him.

"You'll do," he said briefly.

At a safe distance a crowd was gathering, a crowd of Earthmen. Grim surveyed them carefully. They were milling back and forth, but no one dared come closer. "Slaves," he grunted, "not a spark left in them." His eyes swept the heavens. Two faint black specks appeared in the blue distance, from the direction of Great New York.

"They're coming for us," he said quietly.

"Let them," crowed the fiery little bantam, "we'll meet them man to man."

He wrenched the tube from the stiffened fingers of the dead guard, swung it exultingly aloft.

"You little fool," Hilary cried sharply, and struck it down again. "We're not waiting for them. That's suicide. Come. I'm afraid it's too late for you to turn back now. You've been seen with us."

He dashed across the moving belts, Grim and Wat, a grotesquely assorted pair, directly behind him.

Passengers, men and women both, scattered at their approach, stark, servile fear smothering their dulled countenances. Cries arose on all sides. "The Magnificents are coming."

The black specks became larger, forming themselves into swift one-man fliers. The three men pelted across the graded conveyors as hard as they could run. No Earthman tried to stop them; one look at their grim faces would have been a most potent dissuader. And fortunately there were no Mercutians within hailing distance other than the rapidly nearing fliers.

They flung themselves off the last slow-moving platform, panting.

"Which way now?" Hilary asked. His quick eye raked the scene for possible hideouts. They were on a smoothly clipped lawn, heaving gently up to a pretty rambling structure, built on an antique design, pleasingly irregular and nestling to the ground as though it were indigenous to the soil. The walls were modern, though, of vita-crystal, which possessed the peculiar property of permitting *all* of the beneficial rays of the sun to penetrate, and yet presented a perfectly opaque appearance to the outside world.

No other hiding place was in sight. The lawn stretched smooth on all sides except for a scattering of trees—poor enough cover. The Mercutians were almost directly overhead now, preparing to swoop.

Our only chance seems to be the house," Hilary answered his own question quietly.

Grim shook his head. "Their search beams can penetrate the vita-crystal walls as though they were transparent glass."

Hilary's heart sank. "Can't help it," he said laconically. "Come on."

The three men broke into a run. It was only a hundred yards, but the Mercutians were coming down fast. They had been seen. A flash as of molten metal gleamed overhead. A blinding ray leaped for the ground, struck viciously a little ahead of the running men. The velvet green grass crisped to ash; the soil underneath scorched.

"Scatter!" Hilary shouted.

Instantly the men spread out. Another blast hissed down at them, so close to Hilary that the heat seared his left side like a red-hot iron. The Mercutians were getting the range. Wat Tyler stopped short with a howl of defiance. He whipped the hand tube he had taken from the dead guard out of his blouse.

"Hide it, you fool!" Hilary yelled back at him. "We don't want them to known we are armed."

Wat obeyed reluctantly. He shook his fist high in the air, and started to run again. It was not an instant too soon, either. A beam slithered down the smoldering air, and the Earth literally boiled under its impact, directly on the spot where Wat had stopped to shake his fist. All about them the terrible rays were slashing now.

But a last desperate burst of speed carried the Earthmen onto a wide enclosed portico, in the old manner. Hilary pounded on the vita-crystal door. It was tightly locked.

Step back a moment," Morgan rumbled.

Hilary obeyed. The big man spat thoughtfully upon his hands, worked his shoulders tentatively. Then he too retreated to the outermost edge of the portico. Above, the crystal suddenly shattered. Sharp-edged fragments showered down upon them. There was little time to waste.

Grim heaved forward in a slanting rush, right shoulder extended. He crashed into the locked door like a runaway train. There was a grinding noise, a smash of crystal, and his shoulder was through, incased in a halo of bright, sharp edges.

The big man staggered back, his shoulder streaming blood from a hundred cuts. His face was pale and drawn.

"Good fellow," Wat yelled, "even though you are an overgrown ox." He darted in behind the man-mountain like a twisting snake. His deft fingers reached in through the shattered crystal, pressed something on the inside. The door slid into its wall pocket with a sound of grinding glass.

Wat burst into the opened room first, Grim right behind him. Hilary brought up the rear, Grim's great bulk blotting out for the moment any view of the interior.

There was a sudden gasp—a girl's voice.

"Wh—What does this mean?" She was tremulous, yet unafraid.

Hilary stopped suddenly as though brought up against a solid wall. His heart pounded madly. That voice—but it was utterly impossible!

Wat answered, gallantly. "Sorry to annoy you, miss, but they're after us. My partner here's wounded."

"Oh, you poor man." There was quick sympathy in the clear tones. "But who is after you?"

A splintering crash resounded outside.

"The Mercutians, as you no doubt hear," the little man responded with faint irony.

The girl gasped again. "Oh my God!"

There was silence. Hilary strained his ears, yet took care to keep hidden behind Grim's huge frame. What would she do now? It seemed to him as if the whole world depended on her reply.

The girl broke the silence. She had come to a swift decision.

"They must not get you. Go upstairs, quietly, into the chamber on the left of the hall. It's my bedroom. Their search beams can not penetrate it; the walls are draped with lead-encrusted curtains. I'll stay down here and try to throw them off the trail."

Hilary's heart recommenced beating. A gush of joy overwhelmed him. The girl had proven herself.

Grim spoke, for the first time.

"You know the penalty of course, for hiding us."

She did not answer directly. "I can't help it. I can't surrender Earthmen to those beasts. Besides"—there was a catch in her voice—"it does not matter much since "

Hilary stepped quietly from behind Grim's overshadowing bulk.

The girl's eyes went wide at the sight of him; her slender white hand flew to her throat. She looked as if she had seen a ghost.

"You—you!" she choked. "Hilary!"

She swayed and would have fallen, had Hilary not jumped to catch her. His heart was beating thickly with excess of emotion. Joan Robbins in his arms again—how he prayed for this moment in the icy reaches of interplanetary space. Yet what was she doing here in Bronxville? Her home had always been atop the windswept Robbins Building in Great New York.

Her hand went softly over his features, as though to assure herself that it was really he.

"Oh, my dear," she whispered brokenly. "I had almost given up all hope. Everyone was certain you were lost—long ago."

Whirrings sounded outside.

"Sorry to break up your reunion," Grim interrupted in his bass rumble, "but the Mercutians have landed on the lawn. They'll be in here right away."

Joan tore herself out of Hilary's arms. Her slim straight figure tautened; her velvet soft eyebrows puckered over deep-lit pools.

"Upstairs quickly, all of you," she cried. "I'll manage them somehow."

Hilary said quietly, "I won't leave you alone with those brutes. You go along up, and I'll remain here." The automatic gleamed in his hand.

"No, no," she panted, "you mustn't. You wouldn't have a chance. Leave it to me." She literally pushed them with her little hands to the stairway. "Go, if you love me."

"The girl's right," Grim said, "there's a chance. If not," he shrugged his shoulders, "we can always come down again."

Outside were heavy thuddings on the portico.

"You in there," a heavy alien voice resounded, "open or we blast our way in." The door had been slid back into position.

There was no room for further argument. Very reluctantly Hilary followed his companions up the winding stairway.

At the top of the stairs an entrance slide showed darker on the left. Wat fumbled for a moment until he found the button. The door whirred open, even as they heard Joan's clear voice below: "Come in, Magnificents!" There was a trampling of feet.

CHAPTER IV
THE KIDNAPPING OF JOAN

The Earthmen moved quickly and quietly into Joan's room. Thin, crinkly draperies of heavy silk impregnated with lead in colloidal solution, covered all the walls, the door itself. But Hilary shot no more than a cursory glance around; he had left the slide slightly ajar; he was listening intently. The gun was in his hand. There were only two bullets in the chambers—all that were left of the thousands of rounds the expedition had started out with. He must not waste them.

The thick rough voice of a Mercutian floated up from below.

"Three Earth slaves came in here. Where did they go?"

"They did," Joan admitted readily. "They frightened me out of my wits. I screamed and they ran through the house and out the back way."

The Mercutian was suspicious.

"Hmm. Funny there's no sign of a struggle here. Nothing is upset."

"They ran out the back way," the girl repeated tonelessly.

"We'll see; but if you are lying...." He said no more, but the pause was significant in its implications.

"I would not lie to the Magnificents."

"Not if you are wise." He seemed to be the leader. He evidently turned to his companion, for there issued a flood of throaty consonants to which the other grunted once. Then the listeners heard his heavy stamping as he walked through the house to the rear. A door whirred; he had gone out.

The remaining Mercutian said suddenly: "He won't find them."

"Why not?" Joan asked, a bit tremulously.

The Mercutian laughed harshly. "Because you lied. You've hid them in the house."

Hilary heard Joan's sudden sharp intake of breath.

"No, no, Magnificent," she cried.

The Mercutian laughed again—a hard cruel laugh. There was no mirth in it.

"All Earthwomen are liars. I know where you hid them. In your bed chamber. The trick is too old already. We may not be able to see through the lead curtains, but we can break down the door. I warned Artok not to permit the use of the lead curtains, but he has a soft streak. He listened to the women's pleadings for privacy. Privacy, pah! A cloak for conspiracies, that's all it comes to. When Gurda returns, we search upstairs and drag out your rats from their hole."

He laughed smugly, pleased with his own cleverness.

"It is not so." Strange how calm Joan sounded. "They are not in the house. Only my dying mother is here. She is bedded upstairs. The doctor ordered absolute quiet. The slightest noise would be fatal."

The Mercutian sneered. "We'll take a look at that dying mother of yours right now."

"You mustn't," the girl panted. "She will die, I tell you."

"And what does it matter to me?"

There was the sound of a struggle, a sharp cry, followed by a dull thud.

Hilary was out through the door like a flash, down the corridor to the head of the stairs with automatic extended. The monster Mercutian was coming heavily up the treads. They saw each other simultaneously.

The Mercutian's pink eyes turned a vicious red; the tube dangling in his hand jerked sharply up. Hilary squeezed the trigger. The gun barked. The Mercutian spun half around with the force of the tearing bullet. The deadly beam from his weapon slithered over the wall, searing a great molten gash in the crystal. He was badly hurt, but he did not fall. Howling with pain and rage, he slewed himself around again, pointed his sun weapon unsteadily upward.

Hilary let him have the other slug. The big body jerked, and fell backward with a crash to the bottom of the stairs, there to lie oddly contorted and still.

There was a thundering rush from the rear of the lower floor, a hoarse throaty cry. Hilary tore down the steps three at a time, Grim and Wat slithering behind him.

The other Mercutian was bending over Joan's semiconscious form, sweeping her into the crook of a huge arm. He shot a startled glance at the down-pouring Earthmen, swerved the girl around, and aimed his tube.

Hilary pulled the trigger as he swerved. There was a sharp click, but no explosion. Hilary cursed and threw himself down. He had forgotten that there were no more bullets. The speeding flash scorched overhead.

Grim and Wat crouched low. Wat's tube, the one he had wrested from the dead guard on the conveyor, was being slowly raised. The Mercutian saw it, shifted the inanimate girl in front of himself, and backed stealthily toward the splintered door.

"Don't shoot," Hilary cried sharply. "You'll kill Joan."

Wat lowered the tube disgustedly. Hilary groaned aloud. If only he had had one more bullet. There was enough of the gigantic body exposed to offer an excellent target to a steel slug without harming Joan, but the sun weapon sent out its beam in a flat spray.

The Mercutian sensed their dilemma as they crouched on the stairs. He laughed unpleasantly as he backed through the doorway, Joan's limp body held straight in front of him.

"Good-by, Earth slaves," he taunted. "I take your pretty Earth maiden with me. In five minutes I return, with others. You cannot escape. Good-by."

He jumped clumsily through the door. The crouching Earthmen heard a click. It had closed behind him.

Hilary and his companions cleared the stairs in almost a single bound. He had snatched the sun-tube out of Wat's hand. Through the splintered slide he saw the Mercutian climbing into his flier, but a great crystal column of the portico intervened. Nevertheless, while Wat fumbled for the button that released the slide, he took a chance. Every split second was precious now. He aimed the weapon, pressed the spring. A white dazzling ray darted fanwise from the orifice. It touched the column, fused it into molten, running glass. But the Mercutian was already in his seat, Joan limp beside him. He was fumbling at the controls.

The door slid open at last. Hilary shot through like a bullet from a rifle. The flier had already taken off on a long slanting rise. A three-fingered hand waved mockingly down at him. Hilary raised his weapon, then lowered it with a groan. The flier was well within range yet, but if he aimed the terrible beam at it, there would be a crash of fused twisted material, and—Joan was in it. What a dilemma! If he didn't shoot, she would be borne away—he dared not think to what horrible fate.

Grim's hand rested lightly on his shoulder as he watched the flier become a faint black speck in the direction of Great New York.

"She was your sweetheart." His gruff voice was oddly gentle.

Hilary brushed a weary hand over his forehead. The Earth, the universe itself, were suddenly dead, meaningless gobs of matter.

"Yes," he said tonelessly. "Five years ago she promised to wait for my return. She kept her word. I found her again—only to lose her."

Grim said quietly: "I too once loved a girl. I joined the last rebellion under Amos Peabody. The Mercutians threatened to seize the wives, sisters, sweethearts of the revolters if they persisted. Many of the men surrendered. I was one of those who refused. When the revolt was over, smothered in flame from their giant sun-tubes, I found that they had made good their threats. My girl was gone, vanished. Two Mercutians had taken her away. She was never found again."

He paused in brooding silence. "They are up to their old tricks again." His eyes were steely blue now. Hilary pressed his hand in silence. They were welded together by a common loss.

Wat Tyler broke in upon them. "If you fellows want to hang around here, I'll be on my way. That Mercutian hyena will be back here with a dozen others just like him in less than no time."

Hilary snapped out of his sorrow. He could not help Joan by having himself captured or killed, nor was it fair to Grim and Wat. They had placed themselves unquestioningly under his leadership. Something else too was growing into burning life in his mind. This was his Earth, his and Grim's and Wat's, and of millions of other normal human beings. The Mercutians were interlopers, brutal conquerors. He would devote his now otherwise meaningless life to driving them off the planet, wiping them out of the solar system. A tall order, yes, but not for nothing had he fought almost single-handed against those other monstrosities on other worlds: Martians, Ganymedans, Saturnians. The Mercutians were no stronger than they. Besides, there was Joan.

"Men," he said crisply, once more the clear-headed commander of his space expedition, "I intend to fight these Mercutian invaders until Earth is free once more, or—I am dead. I have no illusions about the magnitude of the job, of its practical hopelessness. But that does not mean that you two have to throw away your lives also. I am a marked man, without any identification tag. You on the other hand, can get away from here, mingle indistinguishably with the hordes of people in Great New York. You would be safe. Our ways part here, if you desire it so." He added hastily, "I would be the last to blame you."

Grim Morgan and Wat Tyler looked at each other, a great giant of a man and an undersized bantam. Yet some electric spark of sympathy seemed to dart between them, these so dissimilar beings.

Wat elected to be the spokesman. His voice rose shrilly, as it always did when he was laboring under stress of excitement or emotion.

"*You* won't blame us," he almost squeaked. "Who asked you? Damn it, haven't we consciences of our own? Are we quitters, yellow-bellied Mercutians to quit a pal? Are we, Grim Morgan? Speak, you big ox."

He wheeled abruptly and shook a small fist high in the air. It barely reached under Grim's nose. The big man looked down at the little gamecock unsmilingly.

"No, Wat Tyler, we are not," he said gravely.

Wat turned to Hilary triumphantly.

"There, you see," he crowed, "we stick together. We'll lick those Mercutian monsters; we'll sweep them into the ocean, into space. And what's more, we'll rescue your girl too." He stopped to catch his breath. Grim was nodding slowly. He had not the little man's exuberance. *His* girl could not be rescued any more, but he could remember.

Hilary's frozen heart warmed into life again. With loyal comrades such as these, even the impossible might be accomplished. Very quietly, without heroics, the three men shook hands. Nothing more, yet they knew that they were bound indissolubly together, as long as there was a gasp of breath in any of them.

Hilary's brain functioned with racing smoothness. In minutes the Mercutians would be back.

"We must find a secure hiding place at once," he said. "Know of any?"

Grim shook his head negatively. "There is none," he spoke slowly. "Their search beams penetrate everything."

"Except lead," Hilary interposed.

"Except lead," he conceded.

"Very well then. We shall have to find a place we can line with lead. In the meantime. I have my space flier up in the Ramapos. If it hasn't been discovered yet, it will be essential to our task. We'll have to get there quickly."

"How?" Wat asked,

"By the conveyors, of course."

"No good," the little man declared. "Mercutian guards will be patrolling them. You have no identification tag. You would be caught."

Hilary considered that. "Suppose you two go on along," he suggested. "Find it and wait for me. I'll manage somehow."

"No," they answered unanimously; "we go together or not at all."

Hilary did not try to argue. He would have replied himself in exactly the same terms. He looked longingly at the abandoned flier of the gray-faced Mercutian, lying cold and still within the house.

"If only we could operate the ship," he said.

Then, characteristically, he dismissed the vain longing and bent to the business in hand. "That means we'll have to make it on foot, and keep under cover all the way. Come on."

As the three men moved rapidly over the great lawn toward the nearest covert, a little wood a quarter of a mile away, the horizon that was Great New York showed silhouetted against the westering sun numerous little black dots. The Mercutians were coming.

CHAPTER V
OUTLAWS OF EARTH

Three days later three footsore, weary, hungry men skulked in the edge of the woods near a little clearing in the Ramapos. For three days they had ducked and dodged and literally burrowed into the ground by day, traveling only at night. Above and around them the noise of pursuit rolled. The Mercutians were persistent.

Speedy one-man fliers patrolled the airways, their search beams casting invisible rays in wide sweeping arcs over the uneven terrain. Wherever they touched, the ground sprang into vivid illumination, crystal clear to depths of ten to fifteen feet. Several times the crystal swath swept breathlessly close to the place where the fugitives crouched in covert. The conveyors carried back and forth armed companies of guards. The Mercutians were making a mighty effort to capture their prey.

But somehow the Earthmen had won through, and eager eyes searched the little glade. Hilary exhaled sharply. The *Vagabond,* stanch and faithful companion of all his travels, rested immovably on the deep green grass. It had escaped the questing eyes of the Mercutians. The travel lanes did not touch this secluded spot.

"So that's your space ship, eh?" said Grim, surveying the tarnished, pitted spheroid with something of awe.

"Yes," said Hilary lovingly as he unlocked the outer port side. A hasty glance around inside showed that nothing had been touched. Everything was orderly, methodical, just as he had left it.

Grim and Wat examined with interest the banked controls, the polarization apparatus that set up repulsion waves and literally kicked the ship out into space away from the planet against which it had been set.

"Time enough to inspect," Hilary warned them. "Never can tell when those damned Mercutians may spy on us."

He set the polarization controls so that the mere pulling of a switch would send the flier careening off into space. He surveyed the apple-pie order of the interior with vast satisfaction.

"Now let them come," he said, "the *Vagabond* can show anything that flies a clean pair of heels. Let's eat."

He dragged an aluminum box out of its locker, opened it to disclose a gray funguslike mass. He cut off huge slices and offered it to his companions.

They looked at it doubtfully.

"Ugh," Wat shuddered violently, "I never saw stuff like that before. It doesn't look good." The little man, they soon discovered, had violent discriminations in food.

"Try it." Hilary assured him. "It's a Martian growth, and delicious. We had to live on the land so to speak, on our journey. Our Earth food gave out long before the finish."

Wat looked at it with manifest distaste, but Grim was already wolfing his portion and making little pleased sounds. Wat bit into a portion gingerly, found it tasted somewhat like truffles, and soon was not far behind in gulping it down.

When their appetites had been appeased, Hilary called a council of war.

"First of all," he told them, "we'll have to find a hideout. That presupposes two things: a place large enough to store the *Vagabond*, and hidden from view, either from the naked eye or their search beams."

"That sounds like a large cavern lined with lead," said Grim.

"Exactly."

"And there are none such in this territory," Grim replied quietly.

"I will not move too far from New York," Hilary spoke with determination; "there is Joan...."

Grim looked blank. There was Joan, of course.

Wat got up suddenly. "I know a place," he said, "within a mile of here, and it's not a cave. Come on; I'll show you. I was a Ranger in the Ramapo Game Preserve in the old days."

Hilary asked no more. The polarization switch made contact, and the *Vagabond* left the Earth with a swift rush. It maneuvered with the ease of an Earth flier. Wat directed him, scanning the rugged tree-clad mountains with eager eye.

"There," he said finally, "set her down right there. Easy."

Hilary saw no break in the uninterrupted line of the mountain, but he followed directions. He had come to have an abounding faith in the little red-haired man.

The space flier eased gently down. Just as it seemed as if it would perforce come to rest upon serrated tree tops, a faint glimmer showed amid the darker green. There was an opening, just barely room for the *Vagabond*.

Hilary jockeyed skilfully through, kept on descending into a narrow cleft in the slope. The walls rose almost perpendicularly on either side. About fifty feet down there was a sharp turn and the gorge angled downward for another fifty feet. When the flier came to rest at the bottom, it was securely hidden in a slanting cleft, some forty feet wide and several hundred long. A mountain brook brawled at one side, assuring plentiful water. The outside world was absolutely invisible. Perpetual twilight reigned; only a pale dir religious light filtered through.

"Just the thing," Wat exulted. "We'll never be found here, no matt how much they search, unless someone actually stumbles into the openin There's almost eighty feet of solid rock above us, and their search bean only penetrate about ten to fifteen."

"Splendid." Hilary said. "Now we've got to get to work."

For two days they toiled incessantly. A rope ladder was fabricated t insure ease of entrance and exit without recourse to the ship. Wat, as the least conspicuous, was delegated to scour the countryside and bring in stores of provisions. The bottom of the gorge was leveled off with infinite labor. Rough wood shelters were erected. Spares and electrical equipment to replace worn parts in the *Vagabond* were also purchased by Wat, in cautious small purchases. It necessitated long trekking through mountain trails, but there was no murmur from him. The search, he reported, seemed to be slackening. Only the routine guards whizzed by on the conveyors, and the usual Mercutian fliers that kept to the regular air lanes.

At last even Hilary was satisfied. He was ready now for the plan that had been slowly forming in his mind during the days of their flight and of work. He was going to attempt a rescue of Joan. She had never left his thoughts once; he was burning with inward anxiety, though his face was a mask to cover his true feelings.

The last evening he sat with the others within one of the wooden shelters. A huge fire of fragrant pine knots blazed up a crude boulder chimney.

"I am going out now to find Joan," he told them quietly.

"When do we start?" asked Wat.

"I am going alone." There was a movement of protest. He checked it at once. "You can understand the reasons. One man can worm his way where three men cannot. It isn't a question of force, of brute strength. Besides, if anything should go wrong, there are still the two of you to carry on—to be the focus of a new revolt. If all of us were caught, there would be no further hope for the Earth."

"It's a hell of a note," Wat grumbled, unconvinced. "There's fighting to be done, and me cooped up here like a sick hen."

"Hilary's right," Grim interposed thoughtfully. "It's a one-man job. We'll have our chance later." He turned on Hilary. "But if anything does happen to you, you understand we won't stay quietly. We'll come—if you are still alive. Promise you will let us know—if you can."

"I'll promise that," Hilary agreed. "There is a way."

He got up and went out of the hut. In a few minutes he was back, holding three small flat disks enmeshed in a spray of fine wires for them to see.

"I've just removed the communication disks from our space suits. Strap them in position on your right shoulder blade, hook the wires—so—and you can talk to me or to each other over distances of one hundred miles. Underneath your clothing they cannot be seen. Should I require your assistance, I'll call, and further, I'll show you both how to run the *Vagabond*, in case...." His voice trailed.

"Yes, yes, of course," Grim interposed hastily, "but you'll be here to run it when the time comes."

"Perhaps," Hilary smiled faintly. Then he leaned forward. "I've gotten a pretty good idea of what's happened on Earth since I went away, but now I need more details. Otherwise I'll run into things that will surprise me, and that might not be so—pleasant."

They told him, interrupting each other, arguing over details, Hilary interposing questions every now and then.

About a year and a half after Hilary's departure into trackless space, a huge flat diskoid came hovering to the ground near Great New York. It carried a party of Mercutians on a friendly exploration, so they said, once communication could be established between Earth linguists and themselves. They were welcomed, made much of. They seemed friendly enough. At their own request they were whirled over the Earth in Earth planes on a tour of inspection.

When they departed, with much protestation of friendship, they assured President Peabody they would return some day, they and others of their race. Just what hidden threat there was in that promise, no one on Earth realized. It was taken at face value.

Just a year later, almost to the day, the by this time familiar diskoid was seen hovering once more over Great New York. The Mercutians were returning. The people of New York suspected nothing. No troops were rushed to the scene to repel invasion; no guns were trained on the space ship. It was just another friendly visit, and hurried preparations were commenced for a rousing welcome on their landing.

What New York did not know was that simultaneously with the appearance of the Mercutian flier over their city, a hundred others were even then hovering over the strategic capitals of the world. The first Mercutian ambassadors had put to good advantage that hurried tour of inspection.

No one was alarmed. Each capital city thought itself signally honored by the reappearance of the lone Mercutian over it. The plan was clever, the timing perfect.

At a signal flashed through the ether, things started happening.

The great diskoids, hovering high in the stratosphere, suddenly blazed into blinding light. To the dazzled onlookers below, a new sun seemed to have been born. A truncated cone of flame leaped downward. The diskoid was the apex, the spreading base all of Great New York. The sheeted brilliance enveloped the doomed city. It was a holocaust. New York became a roaring furnace. Stone and steel heated to incandescence.

The affrighted people had no chance for their lives. Like moths in a flame they died on the streets, in the ovens of their homes, in the steaming rivers into which they had thrown themselves to escape the awful heat. There were few survivors, only those who happened to be inside the giant skyscrapers, protected by many thicknesses of crystal and steel.

As Great New York went, so went a hundred other cities. The Earth was caught unawares, but the governments, the people, responded nobly. Troops were mobilized hurriedly, preparations rushed for warfare.

But the Earthmen did not have a chance. The great sinister diskoids moved methodically over the Earth, high in the stratosphere, where the futile Earth planes could not reach them, and sent the terrible blaze of destruction down unerringly upon armies, cities, towns.

It was over soon. One after another, the Earth governments capitulated. America was the last—old Amos Peabody vowed he would rather go down to utter destruction than yield—but he was out-voted in Council. It was pure slaughter otherwise, without a chance to fight back.

At once the Mercutians set up their government. The Earth was turned into a colony. The leader of the invaders, the son of the Mercutian emperor, became Viceroy, with absolute powers. Sooner or later, it was their intention to transport the entire Mercutian race to the Earth, and make it their permanent home. Mercury was not an ideal place to live on; in the restricted area around the poles where life was possible, terrific storms alternated with furnace droughts, to which the hottest part of the Sahara was an Arctic paradise. No wonder the first Mercutian expedition had broached the subject of Earth as an easy conquest when they returned.

The Mercutians treated the Earth people as slaves. Their rule was brutal and arrogant in the extreme. The Earth people revolted, under the leadership of Amos Peabody. Weaponless, except for small hidden stores of rifles and revolvers—the Mercutians had cannily disarmed their slaves—they fought desperately with axes, knives, clubs, anything, against the overlords.

The result could have been expected. The rebellion was smothered in blood and fire. The bravest of the Earthmen died in battle, or were executed afterwards. The slaves, the weaklings, were left. Old Amos Peabody was treated as Hilary had seen. He was exhibited in city after city as a public warning.

Hilary's blood was boiling as the terrible narration went on and on. But his face was calm, immovable.

"How do the diskoids operate?" he asked.

"Something like the sun rays on the one-man fliers," Grim told him, "only vastly more powerful. They are not limited in range, for one thing. It took only one, fifty miles up in the stratosphere, to destroy all New York. I saw the one that first spied on the Earth. It was about five hundred feet in diameter, made of the same vitreous material, and shaped like a huge lens. No doubt, besides being a space ship, it is just that. The sun's light flashes through it, is rearranged into terrible burning rays, and sears all in its path."

"Hm'm!" Hilary meditated. "So everything the Mercutians have in the way of weapons and armament depends directly on the sun's rays."

"Yes," Grim agreed. "After all, you must remember that with Mercury exposed as it is to the fierce heat of the sun, it would be only natural for them to develop weapons that utilized its rays."

"Then the tubes and the fliers cannot operate at night?"

"Yes, because then they receive the reflected waves from the diskoids that are stationed out in space, in eternal sunlight."

Hilary considered this a moment.

"Where do you think it possible Joan was taken?" he changed the subject abruptly.

"It is hard to say," Grim answered slowly. "But your best chance would be with the Viceroy himself. There have been rumors—when pretty girls disappear."

Hilary's jaw set hard.

"I think I'll interview His Mercutian Magnificence," he said. "Where are his quarters?"

"The Robbins Building."

"Good Lord, that's Joan's...." So that was why Joan was up in the Bronxville suburb. "What happened to her father, Martin Robbins?"

"Executed after the revolt," Wat interposed. "Your girl must have escaped, otherwise she'd have been treated then like the other girls whose relatives had fought."

Hilary smiled unaccountably, the first smile since Joan had been taken. He knew the Robbins Building well; he had been a frequent visitor there in the old days. There were surprises in store for His Nibs the Mercutian....

CHAPTER VI
MUTTERINGS OF REVOLT

The next morning, as dawn burst over the mountain tops, he started on his perilous mission. But no one who knew Hilary Grendon would have recognized him in the meek, shambling, slightly bent Earth slave who climbed the last rung of the rope ladder out of the hidden gorge.

He had changed his clothes for an old, space-worn suit that one of his former comrades could never have any further use for. The skilful application of wood ash and powdered charcoal to the hollows around the eyes, the pits beneath the cheekbones, gave him a gaunt, careworn appearance, suitable to an Earthman too brow-beaten to dream of defying his overlords.

Wat, who had artistically applied the make-up, viewed his handiwork with admiration. "You'll do," he grinned. "The way you look, even a little fellow like me would be perfectly safe in spitting upon you."

Before he went, he explained the mechanism of the *Vagabond* thoroughly to his friends. Finally they nodded; they would know how to work the controls.

There was the question of weapons. The captured sun-tube was out of the question; it could not be secreted beneath the dark-blue blouse. Hilary fondled his automatic wistfully.

"If only I had some bullets," he sighed.

"Hell, man, I know where you can get plenty," said Wat. There was a hidden cache, not far from where they were, stored against the day. There were still some brave spirits left on Earth who hoped and plotted. Wat had been one of them. Hilary's spirits rose immeasurably. With his gun loaded he could face the whole Mercutian planet.

Hilary made the return journey to Great New York in an hour. He wormed his way carefully to the nearest conveyor, and made his way openly to the express platform, secure in his disguise.

There was an air of unrest, of tension in the air. The Earth passengers no longer sat dully, apathetically, as they were whizzed along. Little groups buzzed together, excited, gesticulating.

Hilary unostentatiously joined one. There was a sudden silence as he sank quietly into his seat, glances of uneasy suspicion. But he looked thoroughly innocuous, and the chief whisperer felt emboldened to resume the thread of his interrupted discourse.

"There *are* men left on Earth," he mouthed secretively to the little circle of heads. "The Mercutians went down like animals—fifteen of them killed, I hear. The whole company of guards retreated in a hurry"—he paused for greater effect, and continued slowly and impressively—"from—three—Earthmen."

Hilary raised his head sharply. They were discussing his exploit, evidently. With exaggerations of course. That was inevitable.

"Yes, sir," the speaker proceeded, "that shows you. These damned Mercutians are not invulnerable. They can be overcome, chased off the Earth. But we've got to be men, not slaves."

High excitement shone in the surrounding faces.

"But we ain't got no weapons," a small, weak-chinned man protested.

The other spat carefully: "No weapons, huh? Man, I could show you—"

A dark, silent man standing uninterestedly next to him jabbed him in the ribs. The orator gulped and stammered: "I—I mean—"

"Psst," someone hissed hurriedly, "the Mercutians."

Three giant Mercutian guards, their sun-tubes at the ready, stumbled heavily down the aisles of the express, sagging with the pull of Earth's gravitation. Their gray, warted faces were black as thunderclouds.

They stopped before the hastily scattered group.

"You heard the orders," the hugest one barked: "no congregating of Earth slaves on the conveyors or elsewhere. Next time you disobey, I'll ray you. You understand?"

"Yes, Magnificent," the weak-chinned man muttered hurriedly.

But the little knot reformed immediately after the guards had passed on.

"Magnificents!" The first speaker spat viciously. "I'd like to wring their necks."

Hilary shifted unobtrusively to another excited cluster. There the same procedure was followed. A quiet-voiced man was talking, lauding the exploit of the three embattled Earthmen, skillfully and subtly enkindling enthusiasm, raising wholesome doubts as to the invulnerability of the hated Mercutians.

Numerous patrols of guards stalked up and down the conveyors, arrogant, manifestly itching for a pretext to ray the conquered. But the Earthmen gave them no opportunity. The groups melted at their approach into meek, vacuous individuals; reformed instantly as they moved on. And there were no informers. The Earthmen had resumed their almost forgotten Earth solidarity in fronting the invaders.

Hilary watched the restless shifting groups with a glow of pride. This was his work, the spark he had kindled was being fanned into a steady blaze. These whisperers, these exhorters, who were they? Members of an underground organization? Possibly. Wat and Grim had both belonged to loose circles, vague and shifting in membership. Possibly they were coalescing now, joining up into a world-wide organization. He hoped so. It would make his task easier, it also helped restore his pride in being an Earthman. He had almost thought that this supine listless race of his was not worth rescuing.

He reached the terminal in Great New York without untoward incident. No one challenged this meek, shabby-looking Earthman. The Mercutians gave him barely a glance; the Earthmen disregarded him when they whispered together. Hilary was content; he was not seeking undue notice.

The terminal was the scene of unwonted activity. The conveyors were disgorging crowds of Earthmen, grim, determined-looking individuals. They scattered purposefully through the various exits of the huge building. Hilary noted with interest that there were no women, no children, on the constantly incoming expresses.

The Mercutians were massing, too. The terminal was crowded with guards. They stalked heavily about, shouldering their Earth slaves rudely out of the way, sending them sprawling with sudden quick shoves. It would take only an untoward word, a false movement, to start a massacre. The Mercutians were deliberately trying to egg them on.

But the Earthmen took the abuse, the physical violence, quietly. They picked themselves up, disappeared through the exits, giving way to new arrivals. Once Hilary caught a gleam of familiar steel in the unbuttoned recess of a man's blouse pocket. He smiled. There were untoward events impending.

But first he must take care of his own private matter. Joan was a captive in the hands of the Mercutian Viceroy. What was his name? Wat had told him. That was it—Artok.

He was out in the street now, a wide vita-crystal paved thoroughfare, one of the many that radiated from the terminal like the spokes of a wheel. On either side was an upflung spray of tall receding towers, dazzling in the sunshine.

It struck Hilary suddenly. There had been bright unclouded skies during the days since his arrival. Only at night had it rained, like clockwork: every night for fifteen minutes immediately after midnight. A light steady shower that ceased as suddenly as it sprang up. It was unusual. This was April in the Spring of 2348 and April was always a month of showery heavens. Had the Mercutians, accustomed to the blazing light of their own planet, deliberately managed some way to create perpetual sunshine on Earth? Very likely, considering the clockwork night showers, no doubt for the purpose of preventing droughts. There was the matter of weapons and power, too. They all depended on the sun.

Hilary took the inside moving platform. It would take him to the Robbins Building. The street was black with people, surging back and forth, restless, ominous.

Mercutians stalked purposefully along, in companies of ten. Their guttural voices were harsh with command. The Earthmen scattered out of their way. Those who were not nimble enough were knocked down, trampled underfoot.

One Earthman, braver than the rest, or more foolish, gave vent to a scream of rage, when a young girl, with whom he was arm in arm, was wrested brutally away. His fist shot out, caught the leering guard flush on his chin.

The Mercutian staggered, then bellowed with rage. His tube flashed upward. The Earthman's eyes opened wide as with wonder, then he collapsed, cut cleanly in half.

There was a full-throated growl from the jammed thoroughfare, a sudden surging forward. But the guards, reinforced by others, had their tubes lifted, ominous, death-dealing. The crowd ebbed back hastily.

Hilary had joined the first rush. His blood pounded in his veins at the unprovoked brutality. For a hasty moment he visioned the commencement of the revolt. But as the mob retreated before the weapons, his brain cooled. The time was not ripe yet. It would be pure slaughter. Besides, there was Joan.

Once more he was the meek, downtrodden slave. He got off the platform, shambled over to the Robbins Building, an imposing pile of vita-crystal. It rose high into the air, overtopping even the great Memorial Tower.

Martin Robbins had been wealthy, very much so. He had been a physicist of world repute, and this building was a monument to his inventive genius. The top floors were devoted to marvelously equipped laboratories. On the roof were the living quarters—dwelling of many rooms surrounded by an alpine garden. All Great New York stretched beneath. In the distance the green waters of the Atlantic dazzled in the sunshine.

Hilary knew the layout well. It had been his second home before.... He put the bitter thoughts determinedly behind him. There was work ahead. The stooped, hollow-cheeked creature shambled aimlessly up to the entrance. It was filled with Mercutian guards.

He edged his way along, hoping to pass through unnoticed.

"Here, you," a burly Mercutian barred his way, "get out of here before I ray you."

Hilary seemed to shrivel together in mortal terror. He turned to slink out again. The guard had him by the shoulder, was propelling him with ungentle paws toward the exit. Hilary let himself be shoved.

A cold curt voice spoke a sharp command:

"What have you there?"

Where had Hilary heard that voice before?

The pushing guard spun him around hastily.

"He was trying to get into the building, Cor Urga," he said respectfully. "These damned Earth slaves are everywhere under foot. It's time we rayed a few to teach them a lesson."

Hilary found himself gazing at the gray saturnine countenance that had burnt itself into his memory. Urga—the Mercutian who had kidnaped Joan! His muscles tensed suddenly for a quick spring, then relaxed. He must play the game.

Urga looked him over carefully, puzzled.

"Strange," he grunted, "I've seen this fellow before, but I cannot remember where."

Hilary was taut. Would he be recognized?

But the Mercutian Cor—in Earth terms, Captain of a Hundred—shook his head finally, and turned away. The disguise had held up.

"All these Earth slaves look alike. This one is a particularly poor specimen. Turn him loose. If he tries to come in again, kill him.

"Get," the guard growled viciously, and sent Hilary sprawling out into the street to the muttering accompaniment of the seething Earth crowds. The temper of the people was rapidly reaching the explosion point.

But Hilary picked himself up, meekly brushed himself off, and melted unostentatiously into the moving crowd. He desired no undue attention.

Strangely enough, there were no Mercutians in sight. Only the surging, growling Earthmen. Hilary felt their mysterious disappearance to be ominous—as though they had been warned by some secret signal. Something terrible was about to happen. He must get to that certain passageway he knew, and quickly. If only it were not guarded.

A cry went up about him, a yell of many voices.

"The Mercutians are coming."

Hilary whirled. Down the street, issuing from the terminal, deployed a full regiment of guards, bowed under the strong pull of the Earth, but formidable enough. Sun-tubes glinted dangerously. A stentorian voice reached him. "Clear the streets, you Earth dogs," it roared. "You're been warned enough. One minute to obey and I'll burn you all down."

A babel of excited voices went up. The crowds farther down, near the advancing Mercutians, melted into a wild scramble. Men trampled each other underfoot in a mad attempt to reach safety before the minute's expiration.

Where Hilary had paused, there was a milling indecisiveness. Men were already quietly edging their way toward adjoining buildings, into side thoroughfares; others were more belligerent.

"Kill the bloody beasts!" a man suddenly screamed near Hilary, drawing a pistol from beneath his blouse. He waved it frantically in the air. There was an ugly surge, a low-throated growl. It needed very little for the mob to get out of hand and hurl itself upon the steadily approaching Mercutian regiment.

Hilary acted swiftly. He caught the man's pistol arm, thrust it down sharply out of sight. A quick wrench, and the gun was in his own hand. The man, wild-eyed, opened his mouth to shout.

"Shut up," Hilary hissed fiercely. "Are you mad? You wouldn't have a chance. They'd ray us all clean out of existence." He thrust the pistol back into the man's blouse. "Wait; our chance will come."

"Oh, my God! Look!" someone screamed.

A command shattered the air; the tubes of the Mercutians uplifted; a blinding sheet of flame blazed solidly down the street. The minute's grace was up.

Even at this distance, the heat scorched and seared. There were many unfortunates caught farther down, men who had had no chance to seek safety in time. They melted in the furnace blast as though they were bits of metal in an electric arc.

"Run for your lives!" the shout went up. All thought of resistance was gone. It was every one for himself. The man with the gun was the first to run. Hilary found himself caught in the mad rush. The Mercutians were pounding along methodically raying in front of them.

Hilary was thrust into a little eddy of men to one side. It swirled and shoved. The entrance of the Pullman Building loomed ahead. The sight of it gave Hilary new vigor. That was his destination. If only he could make it.

He straightened out of his stoop, squared his shoulders. The next instant a human battering ram crashed through the twirling, yelling mob. Head down, right shoulder and elbow working in unison, a path magically opened where no path had been before. Every second was precious now. The heat of the tubes was engulfing him in waves, raising little blisters on the unprotected skin.

Hilary plunged into the open entrance of the Pullman Building. It was packed with humanity, struggling for the lift platforms, to take them to the upper stories, out of reach of the awful rays. Hilary was thankful for that. His destination was beneath, in the sub-levels. A moving escalator led downward. It was deserted.

A fierce, wild screaming arose outside, screams that gurgled and died horribly. Hilary felt sick inside. The full blast of the rays had reached the milling crowd. It would be a hideous and merciless slaughter.

Hilary's gray eyes burned, his lips set in a straight, hard line. The beasts would pay for this. He shot down the escalator at full speed. A spray of passageways met him. He did not hesitate. He chose the one farthest to the left and dashed along its winding length until he came to a dead end. The vita-crystal gleamed blankly back at him.

But Hilary knew what he was doing. Long ago Martin Robbins had told him of the secret connection between the two adjoining buildings. A passageway that led between the outer and inner shells of crystal walls; lifts that shot smoothly to the laboratories and pent-apartments on the roofs of the two structures. For Simeon Pullman had been a close friend of Robbins; a fellow physicist, in fact. They interchanged theories, results of experiments, and found this swift connection most convenient.

Both men were dead now—Pullman as the result of a premature explosion, and Robbins, executed by the Mercutians. But the secret passageway remained.

Hilary pressed the secret spring he knew of. A gleaming oblong of crystal slid silently open. He went in without hesitation and the slide closed with a little whir behind him.

A low tunnel confronted him, just barely high enough for him to move without stooping. The walls here were of burnished metal, glowing with impregnated cold-light. It was empty, silent. Evidently it had been undisturbed for years. The Mercutians had not discovered this secret way then.

The tunnel slanted downward for several hundred yards, then turned sharply upward until a vita-crystal wall barred the way. Hilary could hear vague sounds from the other side. He was in the Robbins Building. He turned to the left, where a shaft stretched upward, completely enclosed by crystal walls. A thin oblong edging showed the platform beneath. He stepped on it, hesitated for a moment. There were two control buttons; one that stopped the lift in the laboratory, the other in the sleeping room that once was Martin Robbins'.

Hilary decided in favor of the penthouse; there was less chance of a present occupant of the room. If there was—he shrugged his shoulders and loosened the automatic in his blouse. He pressed the button.

The platform shot smoothly upward, up, up, thrusting a thousand feet up. At length it came to a gliding halt. Hilary knew he was on the roof now, in the interior of the wall making one side of the sleep-apartment. The vita-crystal gleamed mockingly opaque at him. If only he could see through; if only he had a Mercutian search beam now. Was there someone in the room on the other side of the wall? He strained his ears to listen, but the crystal was pretty much sound-proof.

Very quietly Hilary drew his gun, broke it, examined the chamber. The six bullets lay snug. He snapped it back in position, held the automatic butt against his side, reached over and pressed the release button.

The slide whirred open. Hilary waited a second, tense, ready to shoot at the slightest sound. His eyes bored through the oblong. Nothing was in sight except the luxurious furnishings he remembered so well; nothing stirred. But his vision was limited to that part of the room framed by the slide. With infinite caution he peered out, his searching gaze flicking swiftly, around the sleep-apartment. It was a man's room with built in divans, automatic sleep-spray, wall rack to hold illuminated book sheets, magnified so as to be read comfortably from a reclining position on the divan—in short, the usual ordered luxuries of a well-furnished sleep-room.

It was empty—but the divan was touseled, certain small things disarranged. Someone used this room. Hilary stepped out, leaving the slide behind him open in case of an enforced retreat. He paused to think. Where could Joan be held prisoner—if, and it was a big if—she were really here. He ran over the possibilities.

The laboratories were out of the question. The great master room then. No doubt Artok, the Viceroy, had installed himself there. It was regally magnificent. That might repay a visit. A bold scheme flashed across his mind. Seize Artok himself, abduct him into the secret passage, and compel him to disclose Joan's whereabouts, give her up. Hilary smiled grimly. Sheerly suicidal, yes, but he was desperate now, and there seemed no other way.

Gun shifted back into his blouse, with his right hand thrust in, on the butt, he glided softly out of the chamber. No one was in sight. The passageway seemed oddly deserted. Possibly the staff had been attracted to the outer rim of the terrace by the commotion below.

At the end of the passageway, facing him, was the master room. Another swift look about, and Hilary was moving down the long corridor, close to the wall, his footfalls deadened by the soft composition rug.

Slowly, very slowly, he pressed the button to release the slide. It slid open at a barely perceptible rate. As the slender crack widened, Hilary, looked in, taking care to keep his body to one side.

CHAPTER VII
IN THE HANDS OF THE MERCUTIANS

A Mercutian was lolling in a reclining chair, his gray, warty face turned half away from Hilary. He was rather undersized for a Mercutian, standing not more than seven feet, and his gray, unwieldy body was heavy and gross as though thickened with good living and debauch. A fleshy three-fingered hand was pounding vehemently on the arm of the chair. His guttural roughened voice came clearly to the listener. He was talking to someone unseen from the angle of the slowly widening slit. He was annoyed.

"For the last time I give you the opportunity," the Mercutian howled—in English. "If you refuse I turn you over to Urga; he wants you."

The crack in the door had widened perceptibly. Hilary's heart gave a tremendous leap. Disclosed to his vision was a figure standing opposite the Mercutian, slim, defiant, proud—Joan.

What unimaginable luck! The automatic leaped like a live thing into, his hand. He crouched, the blood pounding in his temples, waiting for the slide to come completely open. He dared not reach over for the button control to shift the speed; the movement might be heard inside.

The path was clear now. Overpower the Mercutian, escape with Joan down the deserted corridors back to the secret passageway, emerge below, return to their hideout in the Ramapos and plan for revolt. It was all as simple as that.

We must have these Earth slaves," the Mercutian continued, unheeding. "They, must be made an example of. They are responsible for the unrest. They have killed Magnificents; and the Earth fools think they can do the same. They will find out their error soon enough. But as long as those three live, so long will the slaves hope, and plot."

"I cannot tell you anything about them," Joan said monotonously. It was evident that this was not the first time she had said so.

"Yes you can," the Mercutian said as softly as his gutturals would permit. "There is one in particular you know a great deal about. Urga told me. A long-lost lover, no?" His gray-ridged countenance contorted into a thick disgusting leer.

"There it something mysterious about him. He has no identification tag; he releases Peabody; seems not to know the penalties. He has a pistol, a forbidden weapon; he dares to kill a Magnificent; he eggs on two others, ordinary Earth slaves to join him; he disappears out of sight, in spite of all search." He was shouting now, pounding the chair arm with complete loss of dignity. "Who is he, where does he come from, where did he go? Answer me?"

The girl faced him boldly.

"You are afraid of him, Viceroy," she challenged. "You fear his example. He has shown what a brave man can do; the Earth people will follow him. The Mercutians are not invulnerable."

"Yes," the Viceroy said heavily. He was talking more to himself. Then he realized his mistake. "No, of course not," he growled hurriedly. "Enough of this. You tell me what I want to know or I call Urga in."

Joan's face went white, but she faced him unflinchingly.

"I do not know where he is, and if I did, I would not tell you."

"Very well then." The Viceroy leaned over to the table.

The slide was completely open now.

"I wouldn't call anyone if I were you."

The Viceroy whirled in his chair at the sound of the calm Earth voice, calm yet deadly in its implications. He found himself staring into the stubby opening of an Earth automatic, a forbidden weapon. The hand that held it was steady, and the gray eyes that bored into his were hard as pebbles.

There was a smothered gasp from Joan. "Hilary."

"Yes; come to take you away." He spoke swiftly. "We have no time to waste, Joan. Is there any binding material in the room?"

"I—I believe there is. Dad always kept odds and ends in the store chest near the bookshelves."

"Go and get it then. We'll truss up his most Mercutian Magnificence—No you don't," Hilary said harshly; "keep your hands in front of you and don't move."

The Viceroy was stealthily reaching for the sun-tube dangling from his belt. He jerked his hand back, a cold sweat beading his forehead. Hilary's finger had compressed on the trigger; the slightest extra pressure meant flaming death.

"That's better," Hilary approved.

"You shall pay for this," howled the Mercutian, finding voice again. "You shall suffer a hundred deaths in one."

"Softly," Hilary grinned. "Just a little while ago you were very anxious to meet me. Now that I'm here you don't seem overmuch pleased." Joan was rummaging frantically in the open chest.

The Viceroy started, his unlidded pink eyes opened wider. But he was careful to keep his hands in plain view.

"You are the Earth dog who killed the Magnificents."

"I wouldn't call names," Hilary advised. "It might be unhealthy. But I am that very individual. And I trust"—he bowed mockingly-"to have more notches on my gun before I am through."

"You—you—shall be taken to Mercury. My father has special places for such as you." Joan was coming now swiftly with lengths of wire, soft thick material for swathing.

"Get me there first," Hilary said indifferently. "Gag him, Joan, so he can't open his ugly mouth any more. Then tie him up, well."

Joan thrust the gag into the thick gash of a mouth, choking off a torrent of imprecations in the guttural Mercutian tongue. Then she proceeded to truss him, expertly, efficiently.

"Good job," Hilary approved. "Now with your kind permission, Most Viceregal Magnificence, we shall go." He bowed mockingly. "Come, Joan."

"Not so fast, Earth slave." A cold saturnine voice resounded like the clang of doom behind him. He whirled, shifting his gun swiftly for a quick shot.

A little gush of heat caught his trigger hand as the index finger contracted desperately. The smarting pain tore the pistol out of his hand. It dropped to the floor, unheeded. Hilary found himself staring into the gross unpleasant face of Urga, a sun-tube trained directly at his midriff.

The Earth slave who tried to slink into the building," Urga said, surprised. "How did he get up here?"

"I don't know," the Viceroy said shortly, working the gag out of his mouth. "Don't stand there like a fool. Untie me." Gratitude was not among the Viceroy's virtues.

Urga's face mottled as he hastened to obey. When Artok stood finally released, he glared heavily at Hilary and Joan. Then slowly a smile broke over his warty features, a smile that boded unutterable things. Hilary waited quietly, ready to seize the slightest opening; Joan pressed wide-eyed against his shoulder.

"Know this Earth dog?" the Viceroy jerked at Hilary.

Urga's glance was puzzled. "I told you I threw him out of the entrance, but even then I felt I had seen him before."

"You have, Cor Urga," the ruler laughed shortly. "This is the one who is responsible for the mutterings of the slaves. He slew your comrade, Gornu."

The captain started, peered into his captive's unflinching countenance.

"He's disguised!" he cried. "Let me kill him, Magnificent." He fingered his sun-tube significantly.

The Viceroy was in high good humor now.

"Not so fast. You would let him off too easy. I have a better scheme. We shall show the mutinous dogs how we treat those who revolt against our will."

A cruel smile broke over Urga.

"I understand, Magnificent. Make a public warning of him like that fool Peabody. Rip out his tongue and his eyes, smash his eardrums, and ride him from city to city, in chains."

"Exactly."

Joan shuddered, convulsively. "No, no," she cried aloud in her terror, "don't do that. I'll tell you everything; I'll do—"

"Joan," Hilary interrupted sharply, "not another word." His arm went around her.

She collapsed against his shoulder, sobbing.

"It is too late for bargains now," the Viceroy shrugged indifferently. "We have the man we wanted. As for the other two, you will tell us where they are hiding anyway."

Urga turned to him expectantly.

"Your Magnificence," he urged respectfully, "you promised me the girl, if—"

"Yes, take her." The Viceroy waved a weary hand. "I don't want her; I have too many as it is."

The captain's face lit up with an unhealthy glow. He approached eagerly to seize his prize. Joan gave a little cry of dismay, and shrank closer to her lover.

Hilary tensed in every muscle. Though it meant instant death, he would not permit that towering brute to lay his clumsy paw on Joan.

Urga reached out to clasp the frightened girl. Hilary seemed to uncoil. His fist shot straight up with all the power of his body behind it. It crashed into the jutting jaw of the Mercutian like a charge of high explosive dynol. For all his height and massive strength, the giant toppled over, thudding heavily against the floor.

For the moment Hilary saw freedom ahead. The sun-tube had fallen from the nerveless fingers. He darted for it with the speed of a striking snake. Even as his fingers curled around the handle, there came a roar from the Viceroy.

"Drop it, or I'll cut you in two."

Hilary knew when he was beaten. Slowly, reluctantly, his fingers uncurled. He arose, to meet the gleaming opening of the Viceroy's weapon, and the surprised stare in back of it.

Urga got up groggily, feeling gingerly the tender point of his jaw. There was unfathomable hatred in his lidless eyes.

The Viceroy chuckled throatily.

"I never thought, Cor Urga, to have seen a puny Earthman, a mere midget, overcome a Mercutian. Especially you, a winner of the prize of strength three times running in the arenas."

Urga flushed darkly.

"It was an unexpected blow; it caught me unawares," he said heatedly. "I'll break the slave in two."

"Try it—without your sun-tube," said Hilary laconically.

The captain made a movement toward him.

"Leave him alone," Artok cried sharply. He seemed to enjoy his Captain's discomfiture. "I have other plans for him. Now go. Take the girl with you. I'll watch this presumptuous Earthling."

Urga advanced with an evil grin. Hilary thrust Joan suddenly behind him; crouching like a cat. He would go down fighting. For all his bulk, the Viceroy wheeled on his flank, raised his weapon.

"One false move, and you are dead carrion," he said coldly. His weapon was raised. Hilary was caught between two fires, exposed to the searing blasts that would issue at the slightest pressure.

Nevertheless he intended to strike. A sudden swerving jump, and he might throttle one before he would be blasted into nothingness. It would be

Urga, he decided grimly. He tensed for the final desperate, suicidal spring. The two Mercutians were watching him like unsheathed hawks.

"Good-by, Joan," he whispered, and his muscles went taut.

Urga paused, his weapon came up sharply. One little pressure, and—

There was a commotion in the outer hall, the sound of padding feet. The four in the master room froze into immobility. Two Mercutian guards stumbled panting into the room. They came to a jerking halt, threw themselves prone upon the floor, arms outstretched in obeisance.

"May we speak, oh Magnificent?" they asked humbly.

"Say your say," the Viceroy said crossly.

They rose to their feet heavily, and one of them spoke.

"The Earth dogs are revolting. The Cors of the outlying districts report that the slaves are massing and are marching on Great New York. They are armed with Earth weapons. The Cor of the Third District reports two men responsible—one is a giant among them, almost as tall as our own kind; and the other a puny red-haired firebrand. The Cor has tried to capture them, but they are elusive. Even the search beams cannot disclose their hiding place."

Hilary's heart gave a great bound. Grim and Wat had not waited then.

The Viceroy's face darkened with anger.

"The filthy scum," he growled; "this morning's lesson was not enough. This time I'll slay, burn, smash until there isn't a single rebel left. I'll fertilize their damned Earth with their own black blood. You, Cor Urga," he snapped, "transmit my orders to the Cors of the Hundreds. They are to mobilize their men at once, and proceed in accordance with instructions known to them as General Order One. All conveyors to be stopped except for troop movements. Every slave found with weapons, or acting suspiciously, to be slain on the spot. Flying patrols to scatter in pairs, observe for concentrations of slaves. Ray any gathering without warning. Inform Cor Algor of the *Tora* (this was the great armed diskoid of the Mercutians that had previously reduced Great New York, Hilary found out afterwards) to resume his station over the city, ready to act when I give the signal."

Even in the conflict of emotions, Hilary marveled at the unhesitating, snapped flow of orders. The Viceroy, in spite of his seeming gross lethargy, was a soldier, and an efficient one to boot.

"Yes, Magnificent." Urga bowed low, and departed, thrusting a malignant glance at Hilary.

The Viceroy thrust off from him his bright yellow robes, wriggled his vast bulk swiftly into a close-fitting dull-gray tunic. To his belt he fastened little round knobs; the sun-tube dangled swankily at one side. He was accoutered for battle.

He seemed to have forgotten the existence of the Earthlings.

"You," he snapped to one of the waiting guards, "go to the laboratory at once; convey my strict orders to Cor Eela that the weather machine must function perfectly. There must be no slip-up—his life will answer for it."

"Yes, oh Magnificence." The guard prostrated himself once more, then departed hastily.

Vast echoes resounded in Hilary's mind. "Weather machine—weather machine," he puzzled, holding Joan the tighter. There was more to this than met the eye. He must think.

The Viceroy turned suddenly, stared at them, fingering his tube.

"I could of course have you killed at once," he thought aloud, "and have no further trouble; but then Urga would be angry." His lidless eyes rested fleetingly on Joan. "And I would lose my public warning to the few Earth dogs who will survive. If it weren't that I needed them to till the fields, and work the machines, I would not leave a single one alive."

He seemed to come to a decision. "You'll wait my return." He spoke sharply to the guard. "Bind them up well. Thrust gags into their mouths." He grimaced. "I can taste mine yet. And remember, if they escape, just turn that sun-tube of yours on yourself. It will be pleasanter for you. Understand?"

"Yes, Magnificent."

The obsequious guard caught hold of Hilary, under the watchful tube of Artok, and proceeded with clumsy weighted fingers to tie him up. Hilary did not resist. An idea was slowly forming in his mind. Joan's turn came next.

When they were trussed so tightly that neither could move, the Viceroy smiled mockingly. "We shall meet again, Earth dogs," he said, and was gone.

CHAPTER VIII
RESCUED

The guard looked at his captives sourly, kicked viciously at Hilary to relieve his feelings. There was fighting to be had outside; Earth slaves to be tortured and slain, and he was out of it—wet nurse to a couple of prisoners.

He growled disconsolately. Through an open slide window giving out on the terrace, a confused roaring, a babel of sounds came filtering through. There was trouble below—fighting already, very likely. The Mercutian glanced back at his bound and gagged prisoners. They were immobile, helpless. He looked guiltily about. The great room was bare, silent. With almost furtive movements he opened the door leading to the terrace, stumbled out, and was leaning over the parapet, absorbed in the spectacle of Great New York below.

Instantly, Hilary lifted his head, exerting to the utmost his muscles. He could just see the guard's back, strained over the side. Hilary relaxed rolled painfully over to Joan. She stared at him wide-eyed.

If only he could make her understand. He must get the gag out of his mouth. Every moment was precious; the guard might return momentarily. He screwed his face into tremendous contortions, wiggled his feet as much as he could, worked his jaws, trying desperately to convey his meaning.

Joan watched him puzzled; trying to follow those strange contortions. Beads of perspiration started on his brow as her face registered blank incomprehension. Just as he was giving up in despair, she grasped the idea. Her face brightened, and her shapely head nodded stiffly.

The trussed-up pair started at once to pivot around on the floor. Fortunately the composition was polished, affording little friction. With infinite pains the maneuver was completed. They lay side by side now. Joan's trim feet close to Hilary's head.

Writhing and contorting, she worked the sharp heel of her foot against the thick wad of the gag in Hilary's mouth, and pushed. It was solidly tied, but it gave a little. Encouraged, she redoubled her efforts, pushing with all the limited force of her bound limbs.

The yielding gag cut cruelly, the sharp heel scraped and gouged into Hilary's cheeks, but he did not mind. He was in a fever of apprehension. If only the guard's interest were held by the events below until he had accomplished what he intended!

At last his mouth was free. The gag had been pushed over his nose. Joan rolled away. She had accomplished the task Hilary had set her, but she was still puzzled. What earthly good would it do him to talk?

She found out almost immediately. He was twisting his head, burrowing with his nose against the blouse over his right shoulder. The open tunic give a bit, and he burrowed painfully, Joan watching with growing fascination, until one of the binding wires stopped further progress. But it seemed far enough, judging from the satisfied illumination in Hilary's eyes.

He spoke, his mouth pressed close against the shoulder blade, his tones queerly muffled, thick.

"Grim Morgan, Wat Tyler, Grim Morgan, Wat Tyler," he whispered over and over again. He could not hear if there was any response; his ears were muffled now by the spread gag. He could not help that.

"Grim Morgan, Wat Tyler," he muttered monotonously, "Hilary Grendon calling. Held prisoner with Joan, top of Robbins Building. Guarded. Urgent you free us. Artok has sent out general death orders. I have plan to stop him. Come, quickly."

Over and over he murmured the message, hoping desperately they would hear him in the communication disks strapped to their shoulders.

"Come quickly," he repeated; and then the guard, tiring of the view below, or the streets having been cleared of rebels, came softly into the room. Hilary's head jerked quickly back, the shoulder of his tunic falling back into position.

"Here, what's this?" the guard growled suspiciously, catching sight of the displaced gag. "How on Mercury did you do that?"

He knelt swiftly, thrust the gag back into position with ungentle paw, kicked the unresisting form in the side to show his displeasure, and rose. Hilary's heart pounded; the guard had not seen the inconspicuous disk under the tunic. He was in an agony of expectation. Had his comrades caught his message? Could they rescue him even if they had? Questions that only time could answer.

The guard was alert now; he did not like that queer removal of the gag. There would be no further chance to unbind themselves. What seemed hours passed as they lay cramped, immobile.

The air grew thick and warm, or was it only his imagination? No, for the guard felt it, too. Then something buzzed, intermittently. One long, two short. It seemed to emanate from a round black button on the sleeve of his gray tunic. A signal!

The guard exclaimed something in guttural Mercutian, rose hastily, and closed the open door and window. He pressed another button, and sheeted lead curtains rolled swiftly over the vita-crystal roof, darkening the room, cutting off the rays of diffused sunlight. Then he seated himself not far from the captives, facing them, grinning savagely. Hilary wondered why.

Again what seemed hours passed. Behind the lead curtain, the room had become definitely, uncomfortably warm. The Earthlings perspired; the atmosphere was literally steaming; and in their cramped limbs, the torture was fast becoming unbearable. Only the Mercutian guard did not seem to mind. He was accustomed to far higher temperatures on the arid planet that was his home.

Just as the prisoners were gasping almost their last gasp, the heat seemed to recede, swiftly. At once the guard rolled back the leaden shade opened the door and window again. His grin was broadly triumphant. Something clutched at Hilary's heart; he understood now. The beastly invaders! He struggled furiously at his bonds, but they did not give. He ceased his efforts, panting.

The moments passed. Hilary was giving up whatever slender hopes he had had. Wat and Grim had not heard, or if they did, they could do nothing. A slow, sullen despair enveloped him.

He was watching the guard. That gray-faced giant turned his head suddenly, jumped up as fast as his lumbering alien weight could move, snatched at his sun-tube.

"Don't move an inch, if you want to live," a deep, slow voice vibrated through the room. A well-remembered voice. Hilary would have laughed aloud his relief, but he was gagged. His comrades had not failed them.

The guard dropped his half-raised weapon sullenly, staring at the intruders in dazed incomprehension. Hilary jerked his head around. Framed in the doorway was Grim—good old Grim—a long-barreled dynol pistol steady in his hand. From behind him there darted a little figure, red-haired, freckled, shrill with delight. An old-fashioned submachine gun, abstracted from some museum, weighed heavily under his arm.

It clattered unheeded to the floor as the bantam dived for Hilary and Joan.

"We came as fast as we could when we got your message," he crowed. "Dropped everything." His nimble fingers were making havoc of the knotted bonds, while his nimbler tongue wagged on. "Boy, we have them on the run. We'll sweep them out into space by the time we're through."

Hilary and Joan were free now. Very painfully they rose to their feet, stamping and pounding their arms to make the sluggish blood circulate again. Wat hopped about in his excitement.

"Here, you little runt," Grim's voice boomed at him, "stop jumping around, and tie up this Mercutian. We have no time to waste."

Wat groaned comically. "See how that big ox orders me around," he proclaimed, but he picked up the wire and in a trice had the guard helpless and glaring.

Hilary had recovered his speech.

"Thanks, boys," he told them simply. "I knew you'd come if it was humanly possible. But how did you manage to get through the Mercutians? The building is honeycombed with them."

Morgan grinned. "We came in the *Vagabond*," he said.

"What," almost yelled Hilary, "you mean—"

"That your ship is resting comfortably outside on the terrace. When little Wat here caught your message in the communication disk, we were busy organizing companies of Earthmen in the hills back of Suffern. As recruits poured in, we'd tell them off in hundreds, appoint officers, see that they had arms, or gave them directions where to find the old caches, and hustled them off. Had to shift our quarters continually, because Mercutian fliers would pick us up with their search-beams, and start raying. Had some close shaves. But when we heard you were caught, we turned over the command to the nearest new officer, hurried to the gorge, and here we are. The *Vagabond* handled beautifully."

"I could take her myself to the Moon," Wat boasted.

"Hadn't we better be going?" Joan asked anxiously.

"There is work first to be done," Hilary, answered grimly. "There's a certain weather machine in the laboratory I want to take a look at."

"Weather machine?" Grim echoed, puzzled.

"Yes. The Viceroy let something slip about it. For some reason it's very important to them that it continues to function. I'm curious."

A gasp from Joan. Surprised, the men turned to her.

"Of course," she said breathlessly. "Father had been working on it for the longest time. It was a machine to control weather. Something to do with broadcasting tremendously high voltages, ionizing the air and causing rain clouds to form or reversing the process and scattering clouds back into thin air. This was the Master Machine. All over the Earth, at spaced distances, were smaller replicas, substations, controlled from this one. He had great hopes of furnishing equable weather to all the Earth. It was just completed, when...." She trailed off.

Grim frowned. "Very interesting, but what is so terribly important about it now?"

"You fool," Hilary exploded, "it's as important as hell. Don't you see? What are the Mercutians' weapons? Sun-tubes, sun-rays from their fliers, tremendous burning disks that are their space-ships. Sun—sun—everything they have depends upon the sun. Take away the sun, and what have they? Nothing but their hideous giant bodies—they are weaponless. Now do you see?" He fairly shouted at him.

Grim's face lit up heavily; Wat was dancing insanely.

"Get hold of the machine, reverse the process. Make it form clouds, great big woolly ones. Start a rain that'll make the Deluge look sick; forty days—a year—and we'll drown them all," Wat cried.

"Exactly," Hilary nodded. "Joan darling, you and Wat get into the *Vagabond*, and wait for us. Grim and I will take care of the laboratory."

"What?" Tyler ejaculated. "Leave me cooped up when there's a fight on. I'm coming."

"So am I," Joan was pale but determined.

"Oh, Lord," Hilary groaned. "Listen to me, please," he said patiently. "Time is precious, and I can't argue. Joan, you would only be a hindrance. I for one would be thinking more of protecting you than fighting. As for you, Wat," he turned to the furious bantam, "I'm sorry, but you'll have to take orders. The *Vagabond* must be guarded. If we're cut off, we're through. And there's Joan."

"Well. If you want to put it that way," Wat grumbled.

"I knew you'd be sensible," Hilary said hurriedly, not giving them a chance to change their minds. "At the slightest alarm, take off. Don't try to rescue us if we don't return. The Earth cause is more important than any individual. If you get caught, too, the revolt will be leaderless; at an end."

The men shook hands gravely. Joan, white-faced, kissed Hilary passionately. "Be careful, my dear."

Then the two men were gone, moving cautiously down the corridor with deadened footfalls. Hilary had retrieved his automatic; Grim had his more modern dynol pistol. The guard had been thrust into a corner, bound, unnoticed.

The laboratory was on the floor below. They trod carefully down the inclined ramp connecting all the floors. The corridors, the ramp, were deserted.

"All out fighting," Hilary whispered. "The revolt must be spreading."

Grim swore. "The idiots. I told them not to start anything until I returned. They'll be wiped out—they weren't ready."

Hilary nodded slowly. He thought of the strange heat while he had been captive. There would be very few Earthmen left alive in Great New York now.

They were at the foot of the ramp now. Just ahead gleamed an open slide. A pale-blue light streamed out at them; in the oblong of the interior they could see moving shapes, weirdly cut off, crossing their field of vision; bright gleaming machines, segments of tremendous tubes flooded with the pale-blue light. And over all was a constant hum, a crackling, a whining of spinning parts. The laboratory!

CHAPTER IX
THE WEATHER MACHINE

The two men flattened themselves against the wall so that they could not be seen by a casual glance from the Mercutians inside the laboratory.

"There are a lot of them," Grim whispered.

"Can't help it," Hilary answered grimly. "Have to take our chances."

"Of course," Grim said simply. There was no backing out.

Silently, with catlike tread, they inched their way forward flat against the wall, keeping out of the blue flood of illumination. The shapes, or rather segments of shapes within, moved about, engrossed in the business at hand, unaware of the creeping death.

The Earthmen had reached their stations unobserved, one on either side of the open slide. Very carefully Hilary protruded his head around the vita-crystal, and ducked back almost instantly. But his quick eye had taken in all the essential details in that momentary vision.

There were about a dozen Mercutians in the long laboratory, and each had a sun-tube dangling from his belt, ready at hand. The laboratory was crowded with apparatus, but what had drawn Hilary's attention was a gigantic gleaming metallic sphere set up prominently in the center of the room. Protruding from it at all angles were great quartz tubes, through which a blue light pulsed and flamed. It was connected by huge cables to a spark-bathed dynamo. Other cables writhed through the translucent ceiling. The weather machine!

Hilary took a firmer grip on his automatic, nodded once to Grim. The two Earthmen stepped simultaneously through the open door.

"Raise your paws high and keep them up." Hilary's voice cracked like a whip through the busy confusion of the laboratory. The Mercutians, scattered as they were, whirled around from their tasks to face two deadly weapons held by two determined-looking men.

There was a chorus of strange guttural oaths, but every hand moved skyward, reluctantly.

Hilary picked out the most blasphemous sounding of the cursers, rightly deeming him the Cor in charge.

"You," he said, "what switches regulate the weather machine?"

The Mercutian Cor was a particularly ugly specimen. The gray warts were gigantic, hiding whatever semblance of manlike features there might have been beneath.

"I'll see you dogs burned to a cinder in the sun first," he growled.

"Keep them covered, Grim," Hilary said sharply. "I'll take care of this fellow personally."

He walked straight across the room for the Cor, eyes blazing, index finger on trigger. The Cor, fear staring out of his lidless eyes, backed slowly away from the approaching death. There was a hushed silence.

"I'll tell, I'll tell!" the Cor screamed, as the relentless weapon almost touched his paunchy stomach.

"I thought you would," Hilary said grimly, not for an instant relaxing the pressure against the trigger. "If you value your worthless hide, you'd better talk, and talk fast. What switch reverses the machine, to bring on rain? If you are wise, you won't try to fool me."

The wretch almost stumbled in his eagerness. "By the gray soil of Mercury I'll tell you the truth." His arm flung up, pointing. "That knob over there controls the—"

Hilary never heard the rest. There was a crash at the other end of the laboratory. One of the Mercutians, tired of keeping his arms high extended, had attempted to rest his huge bulk against a laboratory table. It went over with a splintering crash of glassware.

Hilary whirled around to face the noise. As he did so, the Cor seized his opportunity. His right arm dropped to his side, jerked up his sun-tube. Hilary heard Grim's warning cry, tried to pivot back again. But Grim beat him to it. The dynol pistol exploded sharply; the flaming pellet caught the Cor square in his side. There was a dull explosion and the Cor was torn violently into bits. He dropped, a mass of shapeless blobs.

But now hell had broken loose. The Mercutians were not cowards. At the moment of the diversion, every one of them had gone for his sun-tube. A flame streaked close to Hilary's head, shivered the opposite wall into molten fragments. He ducked behind a table and fired. A Mercutian threw up his hands, staggered and pitched forward heavily. Grim's dynol bullets whined in their passage, spattered the laboratory with flying blobs of flesh. They did terrible execution. Hilary's automatic spat its leaden hail.

But the Mercutians were entrenched now behind tables, machinery, whatever cover they could find. The beams from half a dozen sun-tubes slithered across the room, burning flaming paths through the overheated air, bringing the very walls down about them. It could not last long. Already Hilary had a nasty burn across one shoulder; there was a streak of red across Grim's forehead as he hid behind the panel of the entrance, whipping his pistol around to fire, and ducking back again. There were too many of the enemy, and overwhelming reinforcements could be expected any moment. The Earthmen's position was desperate.

Through it all the great weather machine hummed and crackled; the tubes were sheets of surging flame. Hilary cursed softly. If only the Cor had completed his sentence before he died. Hilary would have chanced a sudden rush forward to reverse it, to bring on a deluge of rain and clouds, even though it meant certain death. The machine seemed to gleam at him mockingly; the hum continued with tantalizing smoothness.

"Look out," Grim's voice came to him sharply. He jerked his head back, just in time. A ray streaked past his ear like a thunderbolt. The heat from it scorched his face.

The Mercutians were stealthily crawling nearer, pushing heavy, tables in front of them as shields. He was almost outflanked now. In another minute he would be exposed.

Hilary thought rapidly. His position was untenable. He would have to run for it. A sudden dash to the door might possibly win through. But the machine! He set his teeth hard. If he could not change the weather, at least he could destroy the infernal thing, stop its grinding out perfect sunshine for the Mercutians.

He lifted his weapon. Off to one side a Mercutian arm advanced cautiously, bringing up a sun-tube. He swung on it and fired. The sun-tube clattered to the floor and the arm jerked back, accompanied by a howl of anguish. Hilary smiled grimly, took careful aim at the metal sphere of the machine. The bullet leaped true for its mark. A little round hole showed—but nothing happened. The infernal machine hummed softly as ever.

He cursed, fired again. Another round hole, and that was all. With increasing viciousness he turned his aim on the quartz tubes, pierced them through and through. Before his very eyes, the quartz seemed to run and melt around the holes, to seal them tight as if he had never shot. The blue flames leaped and surged mockingly. The Mercutians were jeering now; raucous calls went up.

Hilary felt a sinking despair. He had failed; would have to run for it now. Small chance to make it, too. Then he heard Grim's deep bass. "Hold it a moment," he said as if he had read his thoughts.

Fascinated, Hilary saw the giant's pistol slowly thrust its long barrel around the edge of the crystal slide. A half dozen rays leaped viciously, for it. But a flaming pellet streaked out of its orifice before it was jerked back.

Hilary could see its red path as it struck the sphere of the machine. The next instant there was a dull explosion and the whole machine disintegrated into a smother of flying fragments. The expanding dynol had done the trick where lead had failed. There would be no more weather control.

But Hilary did not pause to see the finish. Even as the machine burst, he was running across the room, bending low. Fragments whizzed by him at a fearful clip; rays crisscrossed all about him.

But somehow he was through. Grim's finger was on the slide button. It closed with a snap behind him, cutting off the pursuing howls of rage.

Silently the two men darted up the ramp to the pent-apartment, dashed into the master bedroom. The Mercutian guard whom they had left securely bound, was gone. The Earthmen looked at each other, a great fear in their eyes. In one bound Hilary was at the door slide, thrusting it open. He tore out upon the open terrace, Grim right behind him.

They looked wildly about. The terrace was empty. There was no sign of the *Vagabond*, or of Joan and Wat. High overhead hovered a great burnished diskoid. Long streamlined Mercutian fliers darted through the air, but nowhere was there a sign of the familiar sphere.

Hilary gripped his companion's arm. "They've been captured, Grim," he choked.

"Nonsense," the giant said gruffly, to hide his own misgivings. "They just took alarm at something and winged off."

"But where is the guard then?"

Grim shook his head. He could not answer that. Despair overwhelmed Hilary. After all he had gone through, to have Joan snatched away from him at the moment of success. It was terrible. Wat too, that freckled-faced bantam.

"I should never have left them alone," he accused himself remorsefully.

"Here," said Grim sharply, "none of that. You did exactly the proper thing. We'll find them yet."

It was a confidence that he did not feel. There was the noise of padding feet up the ramp. The Mercutians were coming, in force.

Grim gripped Hilary by the shoulder, shook him vigorously. "They're coming. We're trapped."

Grendon snapped out of the lethargy into which he had sunk, face drawn and gray.

"No. There is a way. Follow me."

The first of the Mercutians pounded heavily into the room when Hilary had thrust Grim into the secret lift. He whirled and fired. The Mercutian coughed and fell forward. Other gray warty faces, furious, thrust from behind their dying comrade. But Hilary was in the lift, pressing the button for full speed down. A darting ray showered them with rounded smoking bits of vita-crystal, but they were dropping headlong through the building.

Ten minutes later they emerged cautiously from the entrance to the Pullman Building. It was deserted, deathly still. The two Earthmen stopped short, horror-struck at what they saw.

The streets were shambles. Hundreds of bodies lay sprawled in tumbling twisted heaps. Earthmen all, with here and there the grotesque huge bulk of a Mercutian who had failed to hear the warning signal. The bodies were scorched, blackened. Raw agony appeared on contorted desperate faces. It was not good to look upon.

"Wh—what has happened?" Grim gasped, his breath coming heavily.

"Just a little pleasantry of the Mercutians," Hilary said bitterly. He looked upward. High overhead hovered a gigantic shape, motionless.

Its great disk, burnished and dazzling in the cloudless sky, seemed to cast a sinister shadow over the city it had destroyed a second time.

"There's the toy that did it," said Hilary. "I felt the heat while I was a captive up in the Robbins Building. You must have flown over after, and missed it."

Grim shook a great brawny fist aloft. His deceptively mild eyes were hard flames now. His face was set in great strong ridges. Hilary had never seen him this way before.

"I'll rip every Mercutian to pieces with my bare hands—shred him into little bits." He meant it too. Hilary shuddered.

Far off down the wide thoroughfare came the glint of weapons, the sight of massed ranks. A Mercutian patrol was shambling along, heavy-gaited.

"Come on, Grim, let's get out of here," said Hilary.

They flattened like shadows against the wall, slunk stealthily through radiating streets. As much as possible they kept their eyes away from the sickening sights, the poor burned bodies of their fellow men. Steadily they headed for the branch local conveyors as being less likely to be under surveillance.

The Ramapos was their destination. Hilary went dully, listlessly. Joan was gone again; this time he could not possibly know where. Every step he took though, seemed to lead him farther away from her. His glazed eye searched the shining skies as he stumbled along. Not a sign anywhere of the *Vagabond*. Only the hateful swift-moving Mercutian fliers.

It was only Grim's insistence that kept him going. The secret gorge was the headquarters of the revolt, he argued. If the fools he had left in charge hadn't thrown their men recklessly on New York against his instructions to join that last foolhardy heroic attack, there was still a chance of salvaging the revolution.

CHAPTER X
BACK TO THE RAMAPOS

It was dark when they reached the first swellings of the Ramapo Range. It was dangerous to try and make their way through tangled brush and mountain trails. All night they camped on the bare ground, sleeping fitfully, cramped cold, shivering. They dared not light a fire; it would draw instant unwelcome attention.

When dawn came, they were on the move, glad to stretch their sodden limbs. Unerringly Grim homed for the invisible cleft. Nothing stirred in the forests, even the birds seemed gone. The fog had lifted, the sun blazed forth in unclouded majesty. The damp on them dried quickly.

But Grim shook his fist at the unwitting orb.

"Damn that weather machine," he growled. "Breaking it seems to have made matters worse. Even the regular midnight shower has stopped. I'd give ten years of my life for the sight of a cloud."

"It will never rain again," Hilary said wearily. "It has forgotten how."

The bright sunny sky seemed a brazen hell to the footsore Earthmen. It mocked and jeered at them with sparkling waves of warmth.

Before them was an unbroken mass of underbrush. The next instant they were on the brink of the chasm.

"They haven't found us yet," said Morgan, surveying the looped end of the rope ladder. They climbed swiftly down the swaying rungs. The rock slanted with them, turned sharply and fell sheer. Below there was a confused murmur, the sound of movement.

A voice came floating up to them, sharp, commanding.

"Stop where you are, you two. You're covered."

"It's Morgan," Grim bellowed, not pausing an instant in his descent.

The next instant he dropped lightly to the floor of the gorge. A moment later Hilary stepped beside him.

Men were crowding about Grim, clean-cut, determined-looking Earthmen. Nothing like the men he had encountered on his first trip on the express conveyor. The bottom of the gorge had all the appearance of a wartime camp.

There were at least a hundred men encamped in the narrow cleft, crowded and crowding. A tall man thrust himself forward, spare, angular.

Welcome, Captain Morgan," he cried. "We had given up all hopes of seeing you again."

"Hello, Waters," said Grim. "Where's Lieutenant Pemberton?"

The other looked shamefaced.

"He's, gone," he muttered. "Took two hundred men with him."

Morgan's face was awful. "Disobeyed orders, did he? Where did he go?"

"To join in the attack on Great New York. Reports came in that the countryside was up in arms, moving to attack the Mercutians. I couldn't hold him. Said you were crazy, never coming back. He went, and two hundred of the boys went with him."

Grim said: "Know what happened?"

Waters shook his head. "Our radio communication went dead yesterday afternoon."

"He's dead," said Grim softly. "The others too."

A groan went up as he described swiftly the holocaust of the day before. "That was why I warned you all to wait. We can't fight them yet. But I'm forgetting...." He turned to Hilary, who had remained quietly aside. "This is Hilary Grendon, your Chief. He's the man who is responsible for the revolt. I told you about him. We all take orders from him hereafter. If anyone can beat the Mercutians, here's your man."

A babel of sound burst about him like a bomb. Men patted him on the back, shook his hand, crowded him until he was almost smothered. It was a rousing reception. The kind Hilary had dreamed of on his return from his tremendous flight through space—and had not received.

For his act of revolt, unwitting as it was, had fired the imaginations of the Earth people, who in their degradation and despair had come to believe the Mercutian overlords invulnerable. It had been the little spark that touched off a far-reaching train of events. In the few days that had elapsed Hilary had become a legendary figure.

The sparkle came back to his eyes, his brain cleared of the fog of hopelessness as he took command. Joan was lost—yes—but there was the Earth to be saved.

His orders crackled. The little gorge became a hive of activity. With Grim and Waters as efficient assistants he soon whipped the tiny company into ordered discipline. Absurdly few to fight the Mercutians, but Hilary counseled patience. They were a nucleus merely, he told them. When the time arrived to fight in the open, the peoples of the Earth would swell their ranks.

To provide against the day, he sent scouts out to filter through the surrounding villages and towns; unarmed, to all seeming meekest of the Earthlings. They stirred the embers of revolt with muted whisperings; they found trustworthy leaders in each community to organize secretly all able-bodied men; they returned with tidings of the outside world, with food and other necessities.

Sometimes they did not return. Then others went out to take their places. It was the fortune of war. Day and night a sentinel was posted in a dugout directly under the overhanging lip of the gorge. It was his duty to warn of impending attack; above all, to rake the sky ceaselessly with a crudely-contrived periscope for signs of gathering clouds, be they no bigger than a handsbreadth.

But the heavens were a brass blaze by day and a glittering mask of stars by night. Weather machine or none, in truth it seemed that it had forgotten to rain.

Hilary was hard put to it to restrain the impatience of his men. Reports drifted in from the scouts. The premature revolt had been crushed in blood and agony. New York was deserted except for the Mercutians. The country round had been ruthlessly rayed; not only had the armed bands of Earthmen been ferreted out and destroyed, but peaceful communities had been wantonly burned into the ground.

Strong reinforcements had been rushed to the Great New York territory from more peaceful sectors of the world. There were three of the terrible diskoids hovering within a radius of one hundred miles, ready to loose their hideous destruction at the slightest sign of disaffection.

But this time the spirit of the Earthmen was not broken. Their gait was springier, their glance more forthright than heretofore. For every one knew that Hilary Grendon, the prime mover, the defier of the Mercutians, had escaped. The invaders sought him ceaselessly, offering huge rewards for knowledge of his whereabouts. But there were no traitors. Even these few who knew would suffer unimaginable tortures rather than reveal him to the enemy.

"Patience," Hilary counseled his little band. "I know it is hard; I have my own scores to even. But we could only bring disaster upon ourselves and the cause of Earth's freedom by premature action. What have we? A handful of men, poorly armed. A few pistols; only, three of which can use the dynol pellets; a little ammunition. The rest of you have knives, axes, pitchforks. Poor enough weapons against the terrible rays of the Mercutians. We must wait."

Someone grumbled. "For what? Until the Mercutians finally trace our hideout and ray us out of existence?"

"We must take that chance," Hilary told him quietly. "Let it but rain, and we move at once."

"It never will," someone averred with profound conviction.

It began to seem so as the days passed, and the sun blazed pitilessly as ever. The brief night showers had ceased completely. That seemed the only effect of the weather machine's destruction. Some of the weaker spirits among the men were for disbanding. They were afraid of eventual discovery; anxious about their families, left to the tender mercies of the outlanders. Hilary argued, dissuaded, but to no effect. They were determined to go. If by the end of the week there was no action, they said, they would leave. It was Wednesday then.

Thursday and Friday passed. No change. On Saturday a scout brought breathless tidings. One of the great diskoids had crashed to the ground from its station fifty miles up in a smother of flame and flying fragments. No one knew what had happened; the Mercutians of course threw a strict censorship about the affair.

But rumors flew on winged whisperings. Some war vessel from space had attacked the Mercutian, brought it down. More diskoids were rushed to New York; there were five now menacing the territory.

Grim looked steadily at Hilary when the news was brought to them. A momentary wild hope flared in his friend's eye that died out quickly.

"I know what you're going to say," said Hilary. "You think it is Wat Tyler and Joan, somehow escaped in the *Vagabond*."

The giant nodded slowly. "Why not?" he challenged.

"It's impossible," muttered the other. "Where could they have been all this time? Surely they would have returned to this place. And you forget that Mercutian guard who was freed. No, my friend, they have been killed, the *Vagabond* seized, and that was the end to that."

Morgan shook his head skeptically.

Saturday was cloudless. Sunday morning the malcontents were to leave, to dribble back quietly to their homes. They were sullen, defiant in the face of the openly expressed scorn of the loyal men, but determined.

"No use getting ourselves killed for nothing," they muttered.

Double sentries were posted that night. A gloom hung over the camp. Hilary went to sleep heavy-hearted. This seemed the end of all his visions. Joan dead, Wat too; no hope of freeing the Earth from its slavery. If only he had the *Vagabond*, he'd take off again for the uncharted reaches of spaces, find some little habitable asteroid, live out the rest of his meaningless life there. With these gloomy thoughts he fell at last into fitful slumber.

He was awakened, hours later by a sudden uproar. The camp was in confusion. Sleepy voices tossed back and forth in inextricable babble. Hilary was on his feet in an instant, instinctively slipping his automatic into his blouse. Grim looked huge at his side, unperturbed.

"What's happened?" Hilary shouted to make himself heard.

"Don't know," grunted the other, "but we'll soon find out."

He pushed massively through the milling crowd of sleep-frightened men like a ship shouldering the waves, Hilary in his wake. One of the sentinels appeared suddenly before them.

"You," spat Hilary, "why aren't you at your post?"

The man saluted automatically and gasped.

"The Mercutians have come."

"What do you mean?" Hilary demanded, as a groan went up.

One of the weak-kneed men, sir," the sentry ejaculated, "wouldn't wait until morning to make his get-away. We found him climbing out. Said it would be dangerous in broad daylight. He was in a terrible funk. We had no orders to stop anyone who wanted to leave, so we just jeered him, and let him go. My comrade leaned out to watch.

"As he hit the ground, he was bathed suddenly in light. The next instant the blackness of the night was split by a sizzling flame. It crisped the poor fellow to a cinder, and sheared the head of my comrade clean off. I caught the body, pulled it back into the dugout, but it was too late.

"I knew what had happened, sir. Some damned Mercutian flying patrol had spotted us with their search beam. I didn't wait for more, but scrambled out of the dugout as fast as I could. Up above I saw a one-man flier slanting down for me. It was a-sparkle, ready for another ray. I came down the ladder in a hurry, I tell you."

The man was panting, white-faced. Someone cried: "It's all over; they'll smother us in now."

Hilary swung around. It would take very little to start a panic.

"Stop that," he said sharply. "Now is no time to play the coward." He turned again to the sentinel.

"A one-man flier, you said?" he reflected aloud.

"Yes, sir," the other answered, "and I'll bet he's calling for help right now."

"That's just what I intend putting a stop to," said Hilary grimly. He shifted his gun to an easier drawing position, swung himself aloft on the ladder. "Take over, Grim, until I come back," he shouted down. "If I don't, send others up to get that Mercutian."

"Come down," Grim yelled after him, alarmed. "I'll go up; you're the leader here."

"That's why it's my job. So long."

The men stared up after the tiny ascending figure, lumps in their throats. They would die gladly for Hilary Grendon now; he was proving himself. Grim fumed and waited. Hilary had disappeared above the angled bend.

CHAPTER XI
DRIVEN FROM COVER

Far overhead, Hilary climbed swiftly. He realised the seriousness of their situation. Let that Mercutian flash his message to Headquarters and there would be a swarm of fliers upon them within an hour's time. They would be caught like rats in a trap, without a chance for their lives.

He gritted his teeth and swung himself up the faster. He turned the bend. There was the dark sky above, faintly spangled with stars. The flier was not in sight. Hilary stifled an imprecation. If he had taken off, they were doomed.

He moved more cautiously now, stepping gingerly from rung to rung up the swaying ladder. The cleft widened; he was near the top. He paused. There was not the slightest sound. But Hilary was taking no chances.

With infinite slowness he raised his head over the matted underbrush that masked the entrance. For the moment he could see nothing in the pitchy blackness. Then a dim shape loomed to one side. From within it there came a tiny hum, intermittent, almost inaudible.

Hilary knew what that was: a transmitter. Even then the fatal message was winging through the ether. He did not hesitate. He lofted to the ground with one quick heave, steadied on his swaying feet as the automatic flashed into his hand.

"Throw up your hands, Mercutian," he shouted at the dimly-perceived bulk. "I have you covered." He tensed, straining his ears for any movement that might locate the hidden foe.

The tiny humming ceased abruptly. There was painful silence.

"Don't try—" Hilary commenced. He stopped, swerved his body suddenly to one side. A red glow had warned him. The hurtling ray scorched past him with a crackling blaze. Hilary was off balance, teetered, and went down with a crash into the thorny underbrush, his automatic exploding into sharp flame.

A hoarse guttural laugh came from the flier. "Got you that time, Earth dog," the invisible Mercutian taunted. There was silence. Another belt crashed from the ship, heaved the ground under its impact. Another and another. Still no break in the silence, no cry.

The Mercutian muttered to himself: "The dog is dead, all right." He peered out cautiously. The underbrush was black, sullenly quiet. Great swaths showed where the rays had swept the Earth. With a hoarse chuckle the grotesque giant climbed over the side of his ship. A search beam swung in his hand. He was in deep shadow. He swung the beam in a short arc. There was nothing, only matted vegetation. There was one thick thorny bush he noted, however, extending its bulk behind the bow of the ship. He stepped out a bit, away from the flier's shadow, and swung his beam directly at it. The invisible ray pierced through the interlacing twigs with ease. It picked out a prone figure, lying with arm extended.

The Mercutian chuckled again, but the chuckle changed almost immediately to a throaty cry of alarm. With a swiftness that went incongruously with his awkward bulk, his free arm dropped for his hand ray. There was a sharp burst of flame, a staccato bark. The Mercutian staggered, swayed with sullen pain-widened eyes, and pitched headlong forward.

The prone figure in the bush leaped up, ran for him. The Mercutian was dead, drilled through the heart. Hilary sheathed his weapon grimly. His task was done. One thing, though. How much of the message had been transmitted? He must know. He vaulted over the side of the flier, fumbled around until he found the receiving apparatus. Then he waited, dreading to hear the silence broken. A minute passed, two minutes, and Hilary breathed a sigh of relief. The message had not gotten through.

Then it came—a tiny sparking, an intermittent hum. Hilary's heart sank with hammering blows. He tried to read the signals, but they were in code, or in the Mercutian tongue, which was just as bad. It was not necessary, though. Headquarters *had* heard; they *knew*.

Hilary did not waste an instant in vain regrets. Within an hour the gorge would be a vicious trap; he must get his men out at once. What then he did not know, nor bother. There was the more immediate problem.

He went down the swinging ladder hand over hand, not pausing for the rungs. Every instant was precious now. His hands scorched, but he did not feel the pain.

His flying body collided thudding with a heavy bulk beneath. There was a grunt, the rope jerked from his hands, and two bodies fell cursing, entangled, to the ground. Luckily it was not far distant. He sprang to his feet, found Grim heaving his bulk up more slowly.

"I was coming up after you," the giant growled. "You were gone too long. That's the thanks I get."

Hilary had no time for idle talk.

"Attention, men," he snapped. "We leave at once. You have five minutes to get your arms, ammunition clips and rations, light marching order."

Without a word they scattered alertly to their tasks. It was the discipline of veterans.

"You didn't get the Mercutian?" Grim was troubled.

"I got him all right," answered his leader laconically, "but too late. His message had gone through."

Five minutes later to the dot, the camp was lined up, accoutered complete. They were silent, tense, but smartly erect. Hilary's flash glowed over them in the dark. Then he nodded approvingly.

"Fine work, men. Up that ladder, one at a time," he said. "Each man counts twenty slowly, one—two—three before he follows. Keep your distance, and move fast."

The first man sprang to the ladder, went up swiftly. Twenty seconds later, the next man's foot was on the bottom rung. Up and up they went, one after the other, each man counting off and climbing. Hilary watched them anxiously.

"Hope we make it," he muttered to Grim. "It'll take all of forty minutes to evacuate, and the Mercutians may be on us by then."

It was almost forty minutes to the dot when Hilary's head emerged from the cleft. He was the last man out. The men were lined up on a level bit, nervous, apprehensive. In spite of the discipline, heads automatically jerked upward, raked the sky for sign of the enemy.

Where to now?—thought Hilary. There were no more hiding places as perfect as the one they had just left. They were forced into the open, easy prey for the first lynx-eyed Mercutian. Sooner or later, they would be discovered, and then.... A last hopeless glance at the mocking stars. Never had man yearned more for rain, oceans and oceans of it.

Hilary roused himself. Whatever of despair he felt did not appear in his staccato orders.

"We march at once, men," he said. "Scatter formation, five paces between. At the signal, take nearest cover, and prepare for action. Forward—"

"Too late." Grim's voice was flat, controlled.

Hilary looked around sharply. "What do you mean?"

"Look." Morgan's hand swept aloft. Through the darkling night, faintly visible in the feeble starlight—there was no moon—were driving shapes, a full score of them converging upon the little band.

One look was sufficient. Mercutian fliers hurrying in response to their fellow's signal. There was no time, no chance to escape.

"Very well, men." Hilary commanded, coldly calm. "Take cover. Do not fire until I give the order."

There was instant scattering. The men dived for whatever poor bit of protection they could find: jutting rocks, tree trunks, thin thorny bushes even.

Grim and Hilary crouched together behind a great boulder.

"How many pistols are there in the crowd?" Hilary asked quietly.

"Not many. Outside of your automatic and my dynol pistol, there are two other dynols and not more than a dozen automatics. If only we had the submachine gun with us, but Wat took it along, and he's gone."

"Not much chance, I'm afraid," said Hilary; "but we'll fight it out. Here they come."

The two men crouched lower. All about them was silence; not even a leaf stirred in the heavy breathlessness.

The driving fliers were easily visible now. Ominous hurtling projectiles, coming to crush out the last vestige of revolt on the conquered planet. On they came, purposefully, directly, knowing their way; a full score, converging in a scream of wind against their bows as they dropped straight for the hidden gorge.

It seemed to the hidden watchers as though they would crash to Earth with the speed of their swoop. But at one hundred feet aloft the fliers braked their headlong flight, hovered motionlessly in echelon formation.

A moment's breathless pause—to the hiding men it seemed eternity—and all the uneven terrain, rocks, trees, bushes, the soil itself, burst into glowing white crystal clearness. The Mercutians had turned on their search beams.

Hilary gazed clear through the rock behind which he crouched as though it were a transparency. All around him he saw the prone bodies of his men, naked to the view of all and sundry.

A hoarse derisive chuckle rasped from above. Hilary sprang to his feet; further attempt at concealment was useless. As he did so, the air seemed to split in two, there was a blinding rending crash. Not ten feet from where he stood, the ground tossed in torture. A man screamed—terribly. The first blow had been struck.

Hilary burned with a cold consuming anger. "Up, men, and fire. Aim forward about three feet back of the prow." That was where the pilot would be.

A scattered burst of cheers answered him. On all sides, like crystal ghosts, the Earthmen rose to their feet. They were fighting men.

Hilary took careful aim at a flier almost directly overhead and fired. He could have sworn he hit it, but nothing happened. Grim's dynol pistol flamed redly nearby. The tracer pellet scorched upward, impacted, against the hull of a flier. There was a faint detonation, and the next instant the air was full of flying fragments.

"Got that one," he said softly.

Hilary was conscious of a faint envy. His automatic seemed like a harmless popgun against that deadly weapon. But he drew another bead and fired again. With bated breath he awaited the result. Nothing. Hilary groaned, made as if to throw the useless gun away, when the flier he had aimed at wabbled, tried to right itself, and crashed in a swift erratic loop.

By now the pitifully few weapons of the Earthmen were popping. Two more of the enemy fliers hurtled to destruction. But as at a given signal, the air above them seemed suddenly to flame destruction. With the noise of a thousand thunderbolts the massed rays struck.

The groaning Earth tossed and heaved in billowing waves to escape its torture. The trees were blazing pyres. It seemed impossible for anything that lives within that area to escape instant destruction.

Hilary felt a wave of blinding heat envelop him, and he was thrown flat to the quaking ground. Frightful cries, screams of agony, came to his dulled ears as from a great distance. He heaved himself up wearily, scorched, smoldering, but otherwise unhurt.

Grim," he whispered through thick cracked lips. "Grim, where are you?"

"Here." Strange how tranquil he sounded. A scarecrow of a figure arose almost at his right from a smoldering bush, a giant clothed in smoking rags. In the strange illumination of the search beams he seemed the wraith of a scarecrow.

"Thank God you're alive," Hilary croaked. "The others...?"

Figures were staggering up from the holocaust about them.

Grim's practised eyes counted. "About fifty left," he said, "just one half."

Hilary's voice rose suddenly, strongly. "Keep on firing, men." Once again his pistol barked defiance.

A faint, ragged cheer answered him. A few guns flamed; there were only a handful left.

"God!" someone cried.

The massed ships above were gleaming faintly. Little shimmering sparkles ran over the hulls. They were going to ray again. Hilary went berserk, screamed strange oaths, fired again and again. Grim fired, more slowly. Two of the enemy ships left the formation, plunged headlong. But the shimmering grew brighter. In seconds the terrible bolts would be loosed. It was the end. The Earthmen knew it. They could not survive a second raying.

Grim shouted. Never before had Hilary heard him raise his voice to that pitch. His great arm was upflung. "Look!" he screamed.

CHAPTER XII
THE VAGABOND

High up, a dark blob against the feeble starlight, something was dropping; dropping with the speed of a plummet, straight for the massed Mercutian fliers. From outer space it seemed to come, a plunging ripping meteor.

A search beam must have swung hurriedly aloft, for it flamed into startling being; a spheroid, compact, purposeful, dropping with breathtaking velocity.

Something seemed to explode in Hilary's brain. A great cry wrenched out of his torn throat.

"The *Vagabond*."

Unbelievable, impossible. Yet he could not be mistaken. The *Vagabond* was coming home again!

By this time the Mercutians had seen it too. It meant suicide, that rushing projectile from outer space, but it would take along with it in the crash of its flight a goodly number of the Mercutian fliers. The Mercutians were no cowards, but death stared them openly in the face.

Instantly, all was in confusion. Forgotten the rebellious Earthmen below, forgotten everything but escape from the down-rushing thunderbolt.

Hilary, staring upward, could visualize the fliers working desperately at their controls. The clustered ships vibrated like a school of frightened fish poised for instant flight. Then they were in motion; scattering, wabbling in the terror of their retreat.

The *Vagabond* hurtled down among them like a hawk among pigeons. Its surface glowed with the speed of its flight. To Hilary's fascinated gaze it seemed as if there would be a terrific smash. But the *Vagabond* came to a screaming, braking halt directly in the center of the milling, scattering Mercutians.

Almost simultaneously the air resounded with staccato bursts. *Ratatat-tat-a-tat.*

"Good little Wat," Grim danced insanely. "He's cutting loose the submachine gun."

Hilary woke from his amazement with a start.

"Shoot, and shoot to kill," he shouted above the turmoil. "Don't let a single one get away."

Automatics spat their leaden hail, dynol pellets flamed redly, and over all resounded the rapid drum fire of the machine gun, pouring steel-jacketed death into the confused ranks of the Mercutians.

The monster invaders had lost their heads. Even then, they could have destroyed the Earthmen with their deadly spreading rays. But the strange apparition from above had demoralized them. No one thought of fighting: flight, safety, were the only thoughts in their minds.

Flier after flier went tailspinning to horrible death while his comrades fled in all directions.

It was soon over. The greater number of the Mercutians were twisted smoldering wrecks. The few who escaped were rapidly diminishing dots in the cold starlight.

Its work finished, the rescuing space flier settled softly to the ground, in the midst of the embattled cheering Earthmen, temporarily gone insane.

The air-lock port yawned, and a slim figure darted out, straight into Hilary's outstretched arms.

"Joan!"

Behind her danced a small red-haired individual, his homely features grinning with delight. Under his arm swung heavily a submachine gun. He disappeared almost immediately into the vast bearlike grip of his gigantic friend. His shrill voice went on unceasingly, but strangely muffled, as Grim hugged him. Finally he extricated himself, ruffled, breathless, but still talking.

"What did I tell you, you big ox?" he shrilled. "We'll chase them off the Earth, sweep 'em out into space."

"Why, you little gamecock," the giant observed affectionately, "I'm beginning to believe you can do it."

"We thought you had gone for good," said Hilary, holding Joan tightly to him as if he feared to lose her again. "What happened to you on the Robbins Building?"

"Can't get rid of us that easily, can he, Joan?" The little man smirked knowingly at the girl. "It was all very simple," he went on. "No sooner had you two left us than we heard the thud of a flier landing on the other end of the roof. The pilot looked out at us startled. We recognized each other simultaneously. It was our old friend—Urga."

Hilary clenched his fist. He had a good many scores to settle with the Cor.

Wat saw his action. "I did my best," he stated apologetically. "I ran for the machine gun. But by that time Urga had shot aloft again. Didn't seem as though he wanted to wait. I heard his whistle shrilling in the air. Fliers came thick as flies."

He spread his hands in a quaint gesture. "What could I do, Hilary?" his voice was appealing. "Any minute I expected to have a ray on us. I couldn't wait for you two, the *Vagabond* would have been a little pile of ashes. Besides, there was Joan. She kicked and struggled: she wanted to stay for you, but I shoved her in the ship, locked the port, and went scooting up like a rocket. You should have seen the Mercutians scatter."

For the first time in his life words seemed to fail him. "You—are—not—angry?" he fumbled, looking for all the world like a bedraggled dog who knows he has been in mischief.

"Angry?" Hilary fairly whooped. "What for? For saving the ship, Joan, all of us? Why, you little bit of pure gameness, you did the only sensible thing."

Wat grinned from ear to ear.

"But why," Grim interrupted, "didn't you have sense enough to come back here, instead of scaring everybody to death?"

Wat turned on him indignantly. "Sure," he squeaked, "and bring all the Mercutians along with me? No sir, I shot straight up into the stratosphere, and headed for the Canadian woods. Felt we'd be safe there."

Hilary looked at him. "I've heard," he said overcasually, "that an accident happened to one of the Mercutian diskoids. Know anything about it?"

The redhead grinned. "I was the accident. I wasn't staying cooped up in the wilderness. Joan and I decided we'd do some scouting before we came back; see what was happening over the rest of the world. We were returning from one of those little expeditions, cruising about fifty miles up, when we almost bumped into the diskoid. We saw them first; we had just come out of the shadow of the Earth; they were in the sun. I let them have it before they had a chance to turn on their rays. The bullets punctured them clean;

must have let out their air. I didn't wait to see; ducked back into the shadow again."

"How did you get here in the nick of time?" asked Hilary suddenly. "A few minutes later and there would have been no rescue."

Wat looked, at him in some surprise.

"Why, we got your signal, of course."

"Signal?" Hilary echoed. "I never—" Then he paused. Morgan was grinning sheepishly, "Here, what do you know about this?" he queried sharply.

The giant's grin widened. "Just a little," he admitted. "I'd been playing around with my transmitter. Used some of the spare equipment we had cached for the *Vagabond*, and stepped up the sending radius to a thousand miles or so."

"We received your call in the woods north of Lake Ontario," Joan interrupted.

Grim nodded, gratified. "I thought it might work," he rumbled. "You see," he explained to Hilary, "ever since I heard about that diskoid, I *knew* that the *Vagabond* was responsible. But you refused to believe it. So I worked in secret, rigging up the apparatus. Didn't want to stir up false hopes. I finished it yesterday. When we were discovered, I started sending."

"It took us just ten minutes over the hour to get here from a standing start," Wat boasted. "We almost burned up the old machine smashing through the air, didn't we Joan?"

She nodded happily from her cozy position in the crook of Hilary's arm.

Hilary looked long and steadily at his friends.

"Well—" he finally began, when someone cried out sharply.

A dark shape shot over the rim of the mountainside, swooped down at them in one fierce lunge. Involuntarily the Earthmen threw themselves flat on the ground to avoid the tremendous rush of its flight. At one hundred feet it banked sharply, a circle of light gleamed, and a long blazing streamer thrust its relentless finger at the prostrate figures of the Earthmen.

There was a blinding flash, a roar. Hilary was on his feet, bullets spitting rapidly. But already the lone Mercutian flier had completed his bank, and was zooming out of range. Hilary watched the flier grow fainter and fainter in the starlit distance. Almost he could hear the far-off hoarse chuckle of its pilot.

Then he turned to survey the damage. The Earthmen were up, growling low heartfelt curses. That one blast had been catastrophic.

There on the ground lay the smoking ruins of the *Vagabond,* beloved companion of his space wanderings. For a moment Hilary gave way to a deep-seated despair. This was the end of all his plannings. He had built high hopes on the *Vagabond* in his carefully laid schemes for overcoming the Mercutians. He stood as one stunned.

Someone cried: "A curse is upon us; let us scatter before it is too late!"

It acted on Hilary like a cold shower, that cry of despair.

"No," his voice resounded strong and vibrant. "We did not need the *Vagabond*. It never was part of my plans." A lie, of course, but most necessary. "That Mercutian saved me the trouble of finding a hiding place for it. Come, let us march. At dawn it rains, I *know* it will."

"You've said that every day since the weather machine was smashed," a voice cried out from the rear.

Hilary paused, thrown off his balance momentarily. Yet a second's hesitation would be fatal. It was Joan who answered for him. She sprang forward, lithe and exalted, her dark eyes flashing even in the dark.

"I'll tell you how he knows. I myself had almost forgotten. Tomorrow is exactly two weeks since the weather machine was destroyed. My father, Martin Robbins, built it. He told me then that its effects were so powerful that they lasted for two weeks, even with the machine turned off. Only positive action could bring an immediate reversal, of weather conditions. *That's* how he knows."

Joan had turned the tide. The waverers turned as one man to Hilary. "Lead on! We follow!"

"Very well," he stated quietly. "We can't remain here. The Mercutians will be back soon in overwhelming force, burning for revenge. We march."

To Joan, in barely audible tones: "Is that true, what you said?"

"I—I think so. I remember Dad mentioned a time limit. I think it was two weeks."

"If it isn't, we're facing a damned unpleasant prospect to-morrow," he said grimly.

CHAPTER XIII
THE LAST BATTLE

Dawn found the little band still struggling over the thick-forested mountains in a desperate attempt to avoid detection. They were footsore, weary, their clothes shredded by innumerable sharp thorns, their eyes bloodshot from lack of sleep. Overhead, the paling sky was already dotted with the fliers of the Mercutians; faint sounds came to them of the clumsy thrashing of enemy patrols as they beat the woods for the fugitives. The Mercutians were putting forth all their resources to seek out and destroy these irritant foci of revolt.

At length Hilary called a halt. They were in a little valley, not far from Bear Mountain. It offered some protection from the searchers. The enclosing hills would mask them, from all but search beams directly overhead.

"It is no use going any farther," he said wearily. "We all need sleep and rest. Sooner or later they'll find us, no matter where we go, and then—" He shrugged his shoulders.

The weary, panting men threw themselves down upon the ground, too tired even to eat. Immediately they were in a drugged sleep. Joan was sleeping too, her face pale drawn, but like a little child's in her slumber. Hilary watched her with a sharp pang in his heart. What would the next few hours bring to her, to all of them?

Nor did Grim and Wat sleep either. The three of them squatted on their heels, silent, as the cold dawn wind swept with a great sigh through the valley.

The stars were paling now, the purple sky was enswathing itself in pearly grays. Something glowed pinkly overhead; and was extinguished almost immediately by the prevailing gray.

Hilary started violently. "Did you see that?"

"See what?" Grim was drunk for lack of sleep.

Hilary was on his feet, peering upward. "I thought I saw—there, there it is again."

The other two were on their feet also, weariness forgotten, heads thrown back.

High overhead, in the overturned cup of the sky, an irregular pink wisp formed before their wondering eyes, and vanished again. But more slowly, than the first time.

"Well?" asked Wat, puzzled.

"A cloud." Hilary's voice was a prayer.

"Hell," said Wat disgustedly. "If that's a cloud I'm a Mercutian. There wouldn't be enough water there to moisten a canary seed."

"And even if there were it wouldn't matter now," said Grim calmly. "We're discovered."

A long slim flier shot athwart the brightening sky, paused suddenly in flight as though jerked by an invisible string. The next instant the valley was illumined by a transparent glow. It enveloped the Earthmen, made crystal figurines of the most solid among them. They seemed like wraiths through which, as in a glass, more could be seen beyond. The solid ground, the rocks, were transparencies floating in an ocean of airy nothingness. A search beam!

The flier hung steady, high overhead, holding them in the dissolving area of his beam. Too high to ray them but also too high for their futile bullets. The Mercutians no longer underrated the fighting abilities of their erstwhile slaves.

"He's sending out messages for help," observed Hilary.

"Let's take it on the run," Wat suggested.

"No good. Where could we run to that his beam couldn't follow?"

"Well, we can only die once," Wat observed cheerfully.

"And take as many Mercutians with us as we can," Grim amended. "That's one lucky thing. Their rays have no greater range than our bullets."

"Except the diskoids," said Hilary. "Here's your chance, Wat, to play with your rattle."

The red head, who had lugged the heavy machine gun all the way with him, patted its snout affectionately. "It plays the devil's tattoo," he said.

More fliers materialized in the by now brighter blue of early morning. The sun was just peeping over the serrated tops of the mountains. But still they did not attack.

"Afraid of us," Wat chuckled. "Bet they'll send to Mercury for the whole damn army before they come for us."

The first shock was over. With the inevitable staring them in the face, the men had achieved something of a gay recklessness. Hilary found some natural recessions under overhanging masses of rocks that would afford protection from the searing power of the rays. To be effective, the fliers would have to land in the valley or fly low, thus exposing themselves to the raking fire of the Earthmen's weapons. Hilary posted his little band skilfully underneath these natural shelters in such a way that they would be able to command the bit of sky from every angle.

The men jerked and fidgeted. The heavens darkened with massed fliers, and still they came. The Mercutians were taking no chances.

"Plenty of guests at our funeral," Wat chuckled, sighting along the barrel of his gun.

Hilary left the jesting to the others. He was watching the skies intently.

Joan slipped her arm through his. "You see something that we don't. What is it?"

He nodded with an intent frown. "There are clouds forming up there. The first I've seen since I came back to this planet. Rain clouds, too, if I know anything about it. Look."

Joan tilted her head backward. Thin scuds of vapor darted across the sky, driven by the morning breeze; dissolved and reformed a little farther on. Tenuous wisps, evanescent, wraithlike. The sun shone steadily, unobscured.

"Those little things," said Joan unbelievingly. "Why, if that's all you're depending on, we're finished."

"Nevertheless they are rain clouds. But *when* the rain will come is another matter. Very likely too late."

Grim came hurriedly over from his post near the entrance to the little valley. His face was placid as ever, but his eyes were worried.

"We are being surrounded," he stated calmly.

Hilary sprang to his feet. "What do you mean?"

"Listen. Do you hear it?"

Far down the overgrown trail they had followed into the valley came the noise of heavy stumbling feet, innumerable feet.

"They are taking no chances," said Grim, his countenance unchanged.

Hilary looked swiftly around. The valley was a cul-de-sac, surrounded on three sides of its narrow oblong by precipitous hills. From the fourth side, the Mercutians were coming—an army, from the sound of them. Overhead were a hundred fliers, and more coming. The trap was sprung!

Hilary's voice rang out. "All men without guns down the valley to repel invaders. Those with guns remain at your positions; watch the fliers. Wat Tyler in command."

With a joyous cry the Earthmen started for the narrow mouth of the valley, all without guns. Gone was the helpless feeling of before; now they could fight too. Axes, spades, pitchforks, sticks and stones even, were their weapons.

Hilary thrust his automatic into Joan's hand. "You use it, dear. I won't need it. Come on, Grim."

Morgan smiled slowly, handed over his dynol pistol without a word to a weaponless man and stalked after his leader. His great hand clutched and unclutched unconsciously. This was what he wanted, hand-to-hand fighting.

By the time they reached the foot of the valley, the noise of the oncoming Mercutians sounded like the rumbling of thunder. Secure in their numbers there was no thought of concealment.

The Earthmen were pitifully few, only thirty of them, and wretchedly armed. Hilary disposed of them up the slope of the hill on either side, set them to loosening jutting boulders. He was in command on one slope. Grim on the other.

In a minute the Mercutians would be upon them. A simultaneous attack, no doubt; the fliers dropping low to loose their deadly rays from above as the land force attacked with their hardly less deadly hand rays.

Hilary shot a last hasty glance aloft. His heart gave a great bound. The thin insubstantial vapors of a little before had solidified, taken on a grosser leaden hue. The sky was a sullen gray, shot through intermittently with the broad flares of a sun valiantly struggling to reassert its long undisputed sway. Little flickers of lightning played around the ragged edges of the clouds.

To the most unobservant it was evident now that a storm was in the making. But might it not be too late? The sun still shone, and as long as its light pierced through, the weapons of the Mercutians held all their deadly potency.

The alien invaders sensed the urgent necessity for quick action, for the fliers were dropping now, hundreds of them, to within range. Hilary heard the shouted orders of the Mercutians Cors, the crashing forward of a mighty host, and then the front of the attack burst out of the trees in an engulfing flood of gigantic unwieldy bodies and gray warty faces.

A quick view of the stout ungainly Viceroy, Artok, another of the coldly saturnine visage of Urga in the front rank, and with a roar of gutturals, the attack was on.

Down from above came a myriad blinding flashes, turning the inclosed valley into an inferno of heat and rocking, boiling, shattered ground. Up the valley shot the massed hand rays of the hundreds as they swept along in close-packed trot.

It seemed as if nothing could exist in that blazing, screaming hell. Hilary, stunned, shaken, scorched, felt as if he were the only one alive. Yet as the front of the attack washed up before him, he did not hesitate. He sprang to his feet, swung the nicely hefted long-handled ax he had picked up, uttered a war whoop that went back to remote ancestors, and flung himself headlong into the boiling mass of Mercutians.

As he did so, he caught a fleeting, comforting glimpse of Grim rising to his full height on the other slope, huge hands raised, and crashing down barehanded, silent, into the ranks of the enemy. A cheer went up, a faint ragged cheer, and other figures popped up out of nowhere and dropped feet first into the fray.

Hilary found himself engulfed in a welter of figures that towered heads above him. His ax swung up and down, bit into something soft and yielding. The Mercutian screamed horribly; blood spouted from his wide-split shoulder. He fell stumbling to his knees, and Hilary stepped into the little open space. That gave him more elbow room. A furious towering monster swung his tube around in the press. Hilary ducked as the sizzling ray sped over his head. There were howls of pain as the spreading beam cut a burning swath through the packed Mercutians.

Thereafter no more tubes were raised. The quarters were too close. It was to be hand-to-hand fighting; thousands of giant Mercutians against a handful of puny Earthmen.

Hilary swung his red-dripping ax in ever-widening circles. At every swing a Mercutian tumbled. A little space opened around him, literally hewn out of living flesh. But with strange fierce cries he threw himself again and again into the wall of bodies. There and there only was salvation possible where the sun-tubes could not be used.

Far over to one side he caught glimpses of bodies in violent upheavings, bodies that thrust explosively to either side as from the sharp prow of an invisible ship. Then a great figure heaved staggering into view, bloody, gashed, great arms encircling Mercutian heads, smashing them together like eggshells, flinging them apart, seizing others. Grim Morgan, berserk with bare hands.

Here and there in his own travail Hilary sighted little foci of struggle, Earthmen with ax and pitchfork and spade battling valiantly in a sea of Mercutians. A swirl, an eddy, and all too often a sudden surge and flowing of gray warty faces, and smooth rippleless heads where an Earthman had gone down, trampled into pulp.

Hilary's first rush with swinging flashing ax had caught the Mercutians unawares. They had relied upon their sun-tubes, and in the mêlée succeeded only in inflicting frightful havoc on their own kind. Now, however, they came for Hilary in a solid mass, huge three-fingered hands flailing, seeking to thrust him down by sheer weight of numbers. He swung and swung again, the ax bit deep, but still they came. His arm grew weary from so much slaughter, it rose more and more slowly, and then it rose no more. The bloody ax was wrenched from his nerveless fingers, and he was down, smothered by innumerable trampling bodies. Over him the tide swirled smooth. Heavy feet kicked and battered at his body, hands reached down to pluck and rip at him.

Feebly he tried to fend them off, but the shodden hoofs smashed him down again, gouged at his unprotected face. He struggled, but soon he would not struggle any more.

From afar came to his dimming ears below, a huge shout that shook the ground. Feet pounded him down into semi-unconsciousness; there was a mighty shuffling to and fro over him, and then the feet were gone. A huge well-remembered hand, caught him, heaved him upright. It was Grim. His face was a wreck, battered out of all semblance, but those blue mild eyes were flaming with an unholy light.

Hilary tottered, and the giant shook him.

"Wake up," he bawled; "they're coming again."

With a great effort Hilary cleared his numbed brain, saw the resurgence of the temporarily beaten herd. His fists clenched automatically.

"Good boy," Grim whooped. "Let's get them."

Then they were engulfed, fighting back to back. Hilary seemed to be fighting in a dream. He never had a clear conception of what happened.

Faces thrust themselves into his own, furious, contorted; his fist went out mechanically, thudded against something soft, and the face disappeared. Hands reached plucking for him; he thrust them off, and swung left and right again.

Once he looked dully upward. The sky was gray slate now, festooned with bellying black. No sign of the sun; not the least ray could pierce. The fliers hung aimless overhead, no sparkle to their hulls. The valley was dark too; the terrible rays had ceased raking it with an inferno of heat.

Just before he lowered his upflung face to smash his fist into another face, something wet blobbed on his forehead. A raindrop? Perhaps, but he was too far gone to care now. Life was an endless series of howling Mercutians to thrust fists into.

A cheer rose high, punctured by quick sharp explosions of sound. Guns. Those few remaining of the fighting Earthmen farther up the valley, no longer menaced by the futile fliers, had come down to help their weaponless brethren. Wat's voice was shrill in the land, yelling, exhorting, screaming. A familiar *rat-a-tat-a-tat* came down the wind. The submachine gun was spitting steel-jacketed death. Where was Joan? Hilary wondered wearily.

A face towered over him, a face he knew. Urga. The Mercutian was no longer impassive; his gray countenance was distorted with hideous hate. "I'll break you in two," he mouthed, and lunged for Hilary.

The Earthman came out of his daze at the sight of the other. Strength seemed to flow back into his weary body. His fist came up, clean with all the power that was left in him. It went home with a soul-satisfying crunch. Urga's gray gash of a mouth seemed to smear slowly over the rest of his face. A wild animal scream burst from him as he sagged. Then a swirl of other Mercutians anxious to get at the Earthman eddied him out of view.

Hilary felt better. Now he could die content. Even with their guns, what could a handful of Earthmen do against the resistless, ever-coming tide of Mercutians, thousands of them?

It was raining now, slowly at first, large scattered drops, then heavier and heavier, until the fogged air was a driving sheet of water.

What of it?—thought Hilary bitterly as he fought and slipped and stumbled in the slimy, bloody muck that was now the ground. The Mercutians' weapons were useless, but they did not need them any more. Sheer numbers would overwhelm the Earthmen.

Then to his amazement something happened. The heavens, long outraged by the artificial repression of the weather machine, kicked over all traces and opened their sluices in earnest. The sky was one vast waterfall. The elements roared and rocked; the valley was knee deep already in a spate of waters.

Hilary splashed and waded after his enemies. But they were going. They staggered and trembled in every shaking limb, heedless now of the Earthmen. They slipped and fell into the flood, and stayed there, motionless under the waters. Like Pharaoh's army they were being drowned before the amazed Earthmen's very eyes.

On their own planet it never rained; there was no water except for carefully hoarded underground lakes. This first taste of real Earth weather was too much for them. They could not withstand the driving rain, the water swirling round their knees. All the strength went out of their shaggy frames, their knees buckled and down they went, helpless, destroyed by a natural phenomenon to which they were unaccustomed. They had actually been smothered by the humidity!

Hilary's voice was strong again. With great shouts, he rallied his men. A pitiful handful; only fifteen of the fifty that had entered the valley. But Joan was alive, her face black with burned skin, otherwise unhurt. Wat's grin rose superior to a mask of raw flesh, and Grim, bleeding from a hundred wounds, was still a tower of strength.

It was a strange sight as they stood almost waist deep in the flood, the storm beating down upon them, hundreds and hundreds of bodies floating, bumping against them.

"We must clinch our victory, men," Hilary shouted above the roar of the elements. "We must go to arouse the Earth, sweep the Mercutians into the oceans while the storm lasts, or all our work will go for naught."

A great cheer went up from the little band, and without resting, without food or sleep, they waded their way out of the valley, into civilization once more, carrying their message, arousing the peoples, gathering to themselves like a tiny snowball rolling down a mountainside, a huge swelling army of jubilant Earthmen, Earthwomen, too, moving in resistless flood down upon New York.

The rest is history. Like a torrent they swept down upon the cowed, weakened Mercutians. Those that did not escape in the great diskoids back to their own torrid, waterless planet were searched out, torn to pieces by the infuriated Earth peoples.

For five days and five nights the storm raged, all over the world. The floodgates were opened; outraged nature was taking her revenge. For five days and five nights the sun was hidden behind bucketing gray skies. And for five days and five nights, Americans, English, Chinese, Zulus, Australians, Russians, Bushmen, Argentinians, animated by a common purpose, rose gleefully and smote the invaders. When the sun finally peeped once more from behind the thick blanket of clouds, not a Mercutian remained. Few had escaped; the rest would never see Mercury again.

"We've won," Joan sighed happily, after it was all over, and was able to nestle once more comfortably in Hilary's arms. "Thanks to you."

"You forget Grim Morgan and Wat Tyler, dear."

"Ye-es, they helped, too," she admitted grudgingly; "but without you, what could they have done?"

Hilary started to protest, but over her crown of shining hair, he saw Grim and Wat watching him, grinning like two monkeys. Wat's thumb was raised to his nose in an immemorial gesture.

"You're right," said Hilary defiantly. "What could they have done?"